# SCIENCE EDUCATION: BEST PRACTICES OF RESEARCH TRAINING FOR STUDENTS UNDER 21

**NATO Science Series**

A series presenting the results of scientific meetings supported under the NATO Science Programme.

The series is published by IOS Press and Springer Science and Business Media in conjunction with the NATO Public Diplomacy Division.

*Sub-Series*

| | |
|---|---|
| I. Life and Behavioural Sciences | IOS Press |
| II. Mathematics, Physics and Chemistry | Springer Science and Business Media |
| III. Computer and Systems Sciences | IOS Press |
| IV. Earth and Environmental Sciences | Springer Science and Business Media |
| V. Science and Technology Policy | IOS Press |

The NATO Science Series continues the series of books published formerly as the NATO ASI Series.

The NATO Science Programme offers support for collaboration in civil science between scientists of countries of the Euro-Atlantic Partnership Council. The types of scientific meeting generally supported are "Advanced Study Institutes" and "Advanced Research Workshops", although other types of meeting are supported from time to time. The NATO Science Series collects together the results of these meetings. The meetings are co-organized by scientists from NATO countries and scientists from NATO's Partner countries – countries of the CIS and Central and Eastern Europe.

**Advanced Study Institutes** are high-level tutorial courses offering in-depth study of latest advances in a field.
**Advanced Research Workshops** are expert meetings aimed at critical assessment of a field, and identification of directions for future action.

As a consequence of the restructuring of the NATO Science Programme in 1999, the NATO Science Series has been re-organized and there are currently five sub-series as noted above. Please consult the following web sites for information on previous volumes published in the series, as well as details of earlier sub-series:

http://www.nato.int/science
http://www.springeronline.nl
http://www.iospress.nl
http://www.wtv-books.de/nato_pco.htm

Series V: Science and Technology Policy - Vol. 47 ISSN: 1387-6708

# Science Education:

## Best Practices of Research Training for Students under 21

Edited by

Péter Csermely

*Semmelweis University, Budapest, Hungary*

Tamás Korcsmáros

*Eötvös Loránd University, Budapest, Hungary*

and

Leon Lederman

*Illinois Mathematics and Science Academy, Aurora, USA*

Amsterdam • Berlin • Oxford • Tokyo • Washington, DC

Published in cooperation with NATO Public Diplomacy Division

Proceedings of the NATO Advanced Research Workshop on Science Education:
Best Practices of Research Training for Students under 21
Budapest, Hungary
1–3 October 2004



ISBN 1-58603-504-5
Library of Congress Control Number: 2005923360

*Publisher*
IOS Press
Nieuwe Hemweg 6B
1013 BG Amsterdam
Netherlands
fax: +31 20 620 3419
e-mail: order@iospress.nl

*Distributor in the UK and Ireland*
IOS Press/Lavis Marketing
73 Lime Walk
Headington
Oxford OX3 7AD
England
fax: +44 1865 750079

*Distributor in the USA and Canada*
IOS Press, Inc.
4502 Rachael Manor Drive
Fairfax, VA 22032
USA
fax: +1 703 323 3668
e-mail: iosbooks@iospress.com



LEGAL NOTICE
The publisher is not responsible for the use which might be made of the following information.

PRINTED IN THE NETHERLANDS

# Contents

**Session V. Successful Practices for Research Training – Central-Eastern Europe**

# Session I

# Introduction – Opening Addresses

*Science Education: Best Practices of Research Training for Students under 21*
*P. Csermely et al. (Eds.)*
*IOS Press, 2005*




**Peter CSERMELY**
*Hungarian Student Research Foundation, Budapest, Hungary*
*csermely@puskin.sote.hu*

It is my special pleasure to welcome all the participant of this NATO Advanced Research Workshop. The story of this Workshop started 2 years ago, when a similar Workshop was held in Visegrad, Hungary. Many participants of that Workshop are sitting in the room. We had a special feeling two years ago: we found our long-sought friends. That was a wonderful experience to all of us and we decided that the co-operation has to be continuous and a follow-up Workshop should be held in two years. And: here we are. I would like to thank to our major sponsor, NATO for that and to its representative, Ragnhild Sohlberg, to make the organization of this sequel possible.

As you might have seen the program, this meeting is a little bit more focused to practical advances and practical knowledge. How to organize, how to construct wonderful programs for students especially at the age between 16 and 20, so around high school, around high school student age? I think the representatives of the best programs in worlds are sitting in this room, so it is my great pleasure to welcome everyone! Thank you for coming.

As I already mentioned, the atmosphere in the last meeting was so enthusiastic that we decided to organize a Network to have our contact in a regular basis. This network started to function in an informal way. This meeting is an excellent opportunity to make our Network formal. This means no major change (we have been working together now for years) but signals that the ties between the collaborating groups are stronger from now than they were in the past.

This meeting is actually not only a NATO sponsored meeting but it also belongs to UNESCO (not with a special sponsorship at this time, unfortunately, but regarding a moral support and a future support for the Network). This is the time when I have to apologize for not repeating the luxury of the meeting two years ago. The reason is simple: we have approximately half of the support we had at that time. You may remember who were there we paid a private jet to every participant and also organized a space shuttle journey as an evening program. We are not able to repeat this now. But I hope despite of this, you will have a good time. The UNESCO sponsorship is also represented by the fact that his meeting is the satellite meeting of the UNESCO-ICSU World Science Forum, which will be in Budapest in 2005.

You can see no politicians sitting in the presidency. This is because there was a change in the Hungarian government two days ago. Some of the politicians lost their seats, and some of them are probably busy to keep it. Our last hope was Dr. Andras Siegler, the vice-president of the National Science and Innovation Office, who received a call yesterday that he must go to Brussels in an urgent matter. Well, this is something new for us. Now Hungary is a member of the European Union, which gives the wonderful pleasure to our officials to go more often to various meetings in Brussels. So I am apologizing not to have any of them here but I think we might survive without them.

There is an even bigger apology I have to convey to you. Leon Lederman, the co-director of this meeting apologizes for not being here. He is a very busy person and unlike our students: not a teenager any longer. (Not in spirit but in physical terms.) I think we all should understand that. He was a great help for getting this whole Workshop assembled, he supported us all the time, and I am very thankful to him for all what he did.

I am very happy to welcome Ragnhild Sohlberg, the representative of NATO who will talk to you in a minute and also Claudina Rodrigues-Pousada from Portugal, who is representing the Federation of the European Biochemical Societies (FEBS). Claudina is currently the president of FEBS. FEBS is one of the major sponsors of the Network and our initiative.

I would like to thank to the organizers of this meeting. To be honest: I did close to nothing to put this occasion together. The real job was done by Laszlo Fazekas, our coordinator and the heart of our organization, who does not have to be introduced. Neither Tamas Korcsmaros, who was here two years ago, and is here now – as always – to help. Szilard Kui, the new secretary of the Network made already an excellent job and will continue it in the Future. Daniel Császár helped us with the transport and Attila Kun will solve our technical problems. Last but not least the radiating and charming personality of Katalin Sulyok, the president of the Hungarian Research Student Association will help our PowerPoint files to shine up. Many thanks and a great applause to all of them!

As I mentioned, this meeting is more focused to the practical knowledge than its predecessor two years ago, but I am happy that there are experts of talented students, and experts of science education here. We again, will have a rich source of various backgrounds and experiences, and will be able to learn from each other a lot. So thank you very much again for coming and I think it is high time to start the real program. It is my special pleasure to welcome here Ragnhild Sohlberg, who will talk about the NATO and EU Science Education programs.

*P. Csermely et al. (Eds.)*
*IOS Press, 2005*


**Ragnhild SOHLBERG**
*NATO Science Division, Norsk Hydro ASA, Oslo, Norway*
*ragnhild.sohlberg@hydro.com*

**Ladies and Gentlemen!**

Due to traffic problems we had driving to Eger, we are starting a little later than planned, but I am awfully happy to be here. I had the pleasure also of attending the previous Advanced Research Workshop (ARW) on the same topic, here in Hungary, in April 2002. I am extremely pleased to see many friends from last time and also to meet new friends.

To arrange a follow-up was an important conclusion of the past ARW. To be introduced to additional initiatives and to evaluate some of the projects that were in a starting phase in 2002, is important due to the scarcity of proven results in an area that is crucial for all of the NATO and the Partner Countries. The three concepts associated with the title of these two ARWs, are important and certainly interrelated: (1) Science education, (2) talent recruitment, and (3) public understanding.

Why am I here? Good question! That is something one should always ask oneself. Most importantly, both my professional and personal interests coincide. First, on the professional level and as already mentioned, I am a member of *NATO's "Security through Science"* program, i.e., I am on the Advisory Panel that proposed that NATO should sponsor this ARW.

Secondly, I am actively involved with the EU Commission for Research through my position as Scientific Secretary for the European Research Advisory Board (EURAB).[1] EURAB is a high level, independent expert group providing strategic advice to the Commission, partly on the last calls of the current framework program, i.e. Framework Program 6 (2002-2006), but mainly looking towards Framework Program 7 (2007-2011). Several of the nations represented here today, have representatives on EURAB.

## The NATO Science Program

The NATO Science Program[2] was founded in 1958[3] with the establishment of the NATO Science Committee, following the recommendations of a Committee on Non-Military Cooperation in NATO. The report of that Committee of "Three Wise Men" (the Foreign Ministers from Canada, Italy and Norway) asserted that progress in the fields of science and technology can be decisive in determining the security of nations and their position in world affairs. The Science Committee immediately recognized that the training of young scientists and engineers was of paramount importance, and introduced a group of support mechanisms that in essence remain today.

[1] http://europa.eu.int/comm/research/eurab/index_en.html
[2] See: http://www.nato.int/science/e/overview.htm
[3] In 1958 NATO had 15 member countries.

Since the early 1990s[4] the NATO Science Program has served a wider scientific community, as also scientists from the earlier 27 Partner countries of the Euro-Atlantic Partnership Council (EAPC) have become eligible for support.[5] 1999 was a landmark year, in that, with the exception of a small number of Fellowships, the Science Program was transformed so that support was devoted to collaboration between Partner Country and NATO Country scientists or to contributing towards research support in Partner Countries.[6] About 10 000 scientists are currently involved in NATO's Science Program each year, as grantees and meeting participants, or as referees and panel members.

Yesterday I participated in another meeting in NATO's Headquarters in Brussels. The Panel of which I am a member, previously named the "Science and Technology Policy and Organization", is now called the *Human and Societal Dynamics Panel* (HSD-Panel). Due to the change in the security environment over the last few years and in NATO's membership this year, i.e. seven of the earlier Partner Countries are now regular members, the civilian part of NATO's organization as well as the Science Committee have reorganized. Some of the Science Panels have been renamed and some have been given a partly new mandate. During yesterday's meeting of the HSD-Panel, we received detailed presentations of the new organization from the administration. In addition to what I am presenting here today, you can consult NATO's web page which has more details.[7]

The Science Program is now called *Science for Security*. In 2004, after the Istanbul NATO's Summit Meeting, a new program was outlined and priority research topics have been indicated. However, to be very concrete is not easy, since under the heading of "security" many themes and topics can be included.

**The European Union and Science Recruitment Challenges**

I was asked also to say something about the challenges facing the EU with regard to the science recruitment.

Goals decided upon at the European Council at its meeting in Lisbon in January 2000, included under the heading: *"Towards a European Research Area (ERA)",* "...to contribute to the creation of better overall framework conditions for research in Europe... and....to become, by 2010, the most competitive and dynamic knowledge-based economy in the world, capable of sustainable economic growth, with more and better jobs and greater social cohesion."

When the European Council met in Barcelona in March 2002, the Ministers decided that "...overall spending on R&D and innovation [be increased, with]...the aim of approaching 3% of GDP by 2010...". In order to satisfy this latter goal, it would be necessary for the European Union to recruit approximately 500 000 new science and technology researchers and 1.2 million research support personnel. Taking into consideration the large age cohorts retiring, the relatively small age cohorts entering higher education, the growing

---

[4] Spain joined in 1982, hence, in the early 1990s, NATO had 16 members.

[5] The Euro-Atlantic Partnership Council (EAPC) was formally set up in 1997 to succeed the North Atlantic Cooperation Council (NACC) established in 1991. Today the EAPC brings together the 26 NATO Allies and its 20 Partners in a forum providing for regular consultation and cooperation. It meets periodically at the level of Ambassadors and Foreign and Defense Ministers.

[6] In 1999 the Czech Republic, Hungary and Poland joined NATO, giving a total of 19 members.

[7] http://www.nato.int/science/index.html

shortage of qualified science teachers, and the lack interest in science and technology (in particular women)- and that it takes years to become a qualified scientist -- this is no minor challenge!

To attempt to deal with this challenge, the European Commission has taken several measures, including the appointment of two high level expert groups, one which delivered its report, *Women in Industrial Research: A Wake up call for European industry*, in Jan. 2003,[8] and another, which delivered its report, *Increasing Human Resources for Science and Technology in Europe*, in April 2004.[9] These reports include several recommendations, and the Commission has established Action Plans. EURAB also set up a working group which provided recommendations, *Increasing the Attractiveness of Science, Engineering & Technology Careers*, in November 2002.[10]

As you well know, the European Union was also enlarged this year, and has after May 2004 25 member countries. Nineteen of the 25 EU members are also members of NATO, hence, it is not surprising that NATO and the EU share some of the same challenges!

**Conclusion**

I would like to thank the dedicated people who have arranged this exciting and interesting workshop, in particular Professor Peter Csermely and his excellent young staff! I am looking very much forward to these two days!!

---

[8] http://europa.eu.int/comm/research/science-society/women/wir/report_en.html
[9] http://europa.eu.int/comm/research/conferences/2004/sciprof/pdf/hlg_report_en.pdf
[10] http://europa.eu.int/comm/research/eurab/pdf/recommendations1.pdf

*Science Education: Best Practices of Research Training for Students under 21*
*P. Csermely et al. (Eds.)*
*IOS Press, 2005*


**Leon LEDERMAN**
*Nobel Laureate*
*Illinois Mathematics and Science Academy, Aurora IL, USA*

This is another greeting to the readers of the latest NATO book. At a time when the growth of science-based technology shows no signs of slowing, we face – as never before – the need for a much greater general awareness of the way our potentialities and our gained behavior are being modified. We see, almost daily, incredible news of the Martian terrain, the moons of Saturn, and the structure of galaxies and quasars. The discovery of Dark Energy is both exciting as the harbinger of new fundamental laws and discouraging as to the vast sea of ignorance. Other frontiers of science beckon us. Perhaps we will discover that the origin and evolution of the Universe must await progress in human consciousness!

The NATO Workshops continue to open doors to a beautiful, but still mysterious Universe.

# Session II

# Environment and Help of Research Training

*Science Education: Best Practices of Research Training for Students under 21*
*P. Csermely et al. (Eds.)*
*IOS Press, 2005*




# WAYS: The Permanent Global Network of Young Researchers

**Márta MACZEL, Mária HARSÁNYI and Zsuzsanna VÁRADI KALMÁR**
*World Academy of Young Scientists,*
*H-1461 Budapest, P.O. Box 372, Hungary*
*ways@sztaki.hu*

**Abstract.** The World Academy of Young Scientists (WAYS) was officially launched at the World Science Forum, Budapest, in November 2003, and is being established under the aegis of UNESCO. WAYS is dedicated to providing an environment for young researchers - principally between 15 and 40 years of age - to carry out quality scientific research and actively participate in science policy and decision-making. The organization nourishes a global community of new scientists by fostering trans-, inter- and multidisciplinary as well as intergenerational partnership and networking. These collaborations increase the exchange of up-to-date information, prepare youth for scientific careers, and promote dialogue between the scientific community and society. Interested researchers and organizations are invited to visit the organization's website (http://www.waysnet.org) or contact the WAYS Secretariat in Budapest (ways@waysnet.org).

## "Children of the WSF"

Young researchers announced the formation of the World Academy of Young Scientists (WAYS) on the closing session of the World Science Forum (Budapest, Hungary, 8-10 November 2003; *http://www.sciforum.hu*). Speaking in the Hungarian Parliament on the World Science Day (Fig.1.), the young scientists described their mission: "to create a permanent global network for the young scientific community that provides regular input into decision-making on science and technology". General Delegate of WAYS, György Pálfi explained the „double objective" of the new worldwide organization: "to provide an environment to carry out quality scientific research and actively participate in science policy and decision-making".

**Figure 1.** Speakers of the World Academy of Young Scientists on the closing session of the World Science Forum on 10 November 2003: A. Adeleye (USA), S. Sawadogo (Burkina Faso), C. Heller (Spain), Gy. Pálfi (Hungary). Chair: J. Hámori (Hungary).

The international interest in the development of a worldwide network of young researchers is clearly reflected in the increasing number of members and supporting scientists and policy-makers. Presently the organization has some 600 members, originating from nearly 70 countries of five continents and representing all disciplines. WAYS also benefits from the support of some very valuable senior scientists, including six Nobel laureates. Besides their scientific merits, these researchers have expressed their enthusiasm for the education of youth. And what young researchers really need today is the support of well-established scientists who are eager to pass on their knowledge to the next generation.

## Objectives

WAYS, developed under the aegis of the UNESCO, promotes both the scientific and social interests of young scientists - principally between 15 and 40 years of age - by searching for the “ways”, both virtual and actual, to the world of science, to science worldwide, within and among different generations of scientists all over the world.

The objective is to make science more attractive for young researchers, and more comprehensible and accessible to everyone. WAYS encourages the career development of young scientists, facilitates interdisciplinary collaboration and networks among scientists of all generations and from all parts of the world, and influences policy-making worldwide. A crucial aim of the new organization is to bridge the gap between north and south. One of the young founders explained the need in terms of his own situation: Serge Sawadogo said, "I'm doing research in malaria, a disease that ravages my country. There are no facilities in Burkina Faso to study immunogenetics, so I conduct my research in France."

The Network is intended to provide a continuous voice to sensitize scientific institutions and decision-makers to the necessity of creating opportunities for young researchers in S&T, and actively involving them in their shaping.

**History and background**

WAYS is being formed as a continuation of the successful satellite event ("International Forum of Young Scientists" or IFYS) of the UNESCO-ICSU World Conference on Science in 1999. Recognizing the new challenges of science at the beginning of the 21st century and problems that have to be faced by young scientists, the IFYS participants decided to establish an organization that would function as a forum to communicate the opinion of young scientists to decision-makers in science policy.

In 2001, at the UNESCO General Conference in Paris, general consensus was achieved about the opening of the IFYS Secretariat in Budapest. The decision was based on Hungary's long tradition of scientific and civil institutions comprising all generations of young scientists (from secondary school through undergraduate and postgraduate studies to postdoctoral level). These young scientist organizations provide a firm background for the creation of an international network.

In 2002, the IFYS Foundation was established by the Association of Hungarian Ph.D. Students. As a legal entity, the IFYS Foundation was charged with advocating on behalf of young scientists and securing financial support to further the mission of the organization. A new name, World Academy of Young Scientists (WAYS), was also proposed. The acronym WAYS has multiple meanings: ways to access science, to encourage scientific cooperation among generations, to international scientific cooperation, to influence science policy and to eliminate barriers to scientific opportunity.

In 2003, due to the Trustee's effective lobby activity, the WAYS Secretariat was fully equipped and opened in Budapest (1 August 2003) with the support of UNESCO, the Hungarian National Commission for UNESCO, the Hungarian Ministry of Education and the Hungarian Academy of Sciences. The Secretariat, located in the Computer and Automation Research Institute of the Hungarian Academy of Sciences, is presently run by two full-time employees. Mr. Balint Magyar, Minister of Education of Hungary already reported on the establishment of the WAYS Secretariat in Budapest at the UNESCO General Conference on 3 October, 2003 in Paris.

The organization made a very successful debut at the World Science Forum, Budapest - 2003 with the participation of 35 young scientists coming from 18 countries, and with the organization of the session 'Forum of Young Scientists'. The event also provided opportunity for discussing the possible structure, operation and activities of WAYS, as well as membership and partnership issues. It was decided that WAYS would develop a structure according to scientific disciplines and geographical areas, support the organization of specialized scientific events, and encourage and promote the international mobility of talented young scholars. Furthermore, an international provisional board was elected, which coordinates workgroup tasks, such as finalization of the WAYS Constitution, development of external relations, coordination of logistical tasks, fundraising and organization of the upcoming 1st General Conference of WAYS. The work of the Secretariat, also acting as the coordinating centre of the organization, is helped by volunteers and advisory members. Presently national WAYS units are also developed with the assistance of local coordinators.

**WAYS… to the world of science, to science worldwide**

The successful debut of WAYS provided basis for numerous media appearances, such as in Science and European Voice. Science's Next Wave presented the global network of young researchers through the portraits of two board members, Marta Maczel and György Pálfi (*http://nextwave.sciencemag.org/cgi/content/full/2004/06/02/2*). Information on WAYS also appeared in newsletters and on websites, and the news on its formation reached five continents.

As a sign of international recognition, the organization got the possibility to organize its own workshop entitled "*Governance of Young Scientist Associations*" in the framework of the MCFA Career Programme during the EuroScience Open Forum, held between 25 and 28 August, 2004 in Stockholm (*http://www.esof2004.org/programme_events/event_detail_mcfa.asp?eventkey=216*). The aim of the workshop was to give young scientists the chance to debate various aspects of the structure and central activities of WAYS, establish new forms of collaboration and partnership, and exchange visions of how to effectively include young scientists in the making of S&T policy.

WAYS, supported by UNESCO, ISESCO and the Moroccan Government, will hold its 1st General Conference in Morocco from 5 to 7 December, 2004. The information package and call for participation reached 190 countries, and the selection of delegates is in process. The "children of the WSF" are also expected back as organizing partners at the World Science Forum – 2005, and in 2006 a joint conference with the International Council for Science Unions is planned.

Meanwhile the organization is seeking the active participation of young scientists from all over the world in its development. Interested researchers and organizations are invited to visit the organization's website (http://www.waysnet.org) or contact the WAYS Secretariat in Budapest (ways@waysnet.org).

*Science Education: Best Practices of Research Training for Students under 21*
*P. Csermely et al. (Eds.)*
*IOS Press, 2005*



# The FEBS Educational and Youth Policy

**Claudina RODRIGUES-POUSADA**
*Genomics and Stress Laboratory,*
*ITQB, Apartado 127, 2781-901 Oeiras Codex, Portugal*
*claudina@itqb.unl.pt*

**Abstract.** The activities of the Federation of the European of Biochemical Societies (FEBS) are described giving a particular emphasis to the program for the young students at the level of their post-graduation.

## 1. Introduction

Founded in 1964, FEBS is one of the European largest organizations in life sciences with approximately 40,000 members representing 36 constituents Societies and 6 Associated Societies. Its goal is to promote and to encourage the interdisciplinary development of life sciences through the support of several areas of basic research such as Biochemistry, Biochemistry, Molecular Biology, Cell Biology, Immunology and Biophysics. The various activities described below have been implemented to respond to the challenges posed by the rapid explosion in the field of the life sciences and to promote science among the young scientists.

## 2. Advanced Courses Committee

This Committee has an annual program of both lecture and practical courses in active areas of Molecular Cell Biology and Biochemistry. The participation is limited to 15 students in the case of a practical course and up to 120 students in either lecture courses or workshops. As FEBS is very eager to collaborate with other organizations, as EMBO and NATO, several of theses courses and workshops can be co-sponsored by these organizations. FEBS Council has also approved a budget (€540,000 in 2002), to support 15 courses (in average 15 per year). This budget includes Youth travel grants to help the participation of young students in the courses held in countries other their own. There is also a sum of about 200,000€ dedicated to organize annually four high-level practical courses in Eastern Europe with the possibility to request special equipment up to a maximum of 20,000€. Proposals to organize advanced courses, workshops or special meetings on topics of high scientific interest are very welcome and should be addressed to the chairman of this committee

**3. Fellowships Committee**

FEBS provides fellowships with different categories: Long-Term, Short-Term and also Summer Fellowships. Fellowships can only be attributed to the members of the FEBS constituent societies. The long-term fellowships are given to students who have finished their PhD and are interested in performing their post-doctoral studies in another FEBS area for a period of two years. This can be extended for one more year depending on the achievements performed by the candidate. The short-term fellowships are given to PhD students or post-doctoral students (having no more than six years of post-doc studies) to spend a short period (not longer than two or, in exceptional cases, three months) and are aimed to have scientific collaboration, advanced training or employing techniques not available at the usual place of work. Summer Fellowships are intended to provide experience to young promising students in an institution within the FEBS area in a country different from that where the applicant studies. The applicants should normally be registered students in a FEBS country should not have yet submitted a doctoral thesis. Another initiative recently implemented is the FEBS Fellowship Follow-up Research Fund. Its objective is to help young scientists who have been recipients of a FEBS Long-Term Fellowship to start work on return to their country of origin. It was also set up a programme designated as Collaborative Experimental scholarships which are specifically addressed to students involved in research for a doctoral thesis in the currently depressed economies of Central and Eastern Europe. Their aim is to support short visits to well-founded laboratories in Western Europe for the purpose of carrying out experimental procedures, which would be impossible in the students' home country. There is also a award for FEBS distinguished Young investigator in order to recognize excellent research conducted by young scientists who have been recipients of a FEBS Long Term fellowship. In the site http://www.febs.org/Activities/Fellowships you have both the necessary guide–lines and the documents to be filled which should be then sent to the respective chairman.

**4. Publications**

FEBS has also launched two main publications European Journal of Biochemistry and the FEBS Letters, respectively. Both journals are widely available both in paper and electronic form, are indexed through the various indexing systems, and are peer reviewed. They publish full length, original papers on fundamental aspects of Biochemistry, Molecular and Cell Biology and Molecular Biophysics. FEBS activities could not be possible without the income provided by these two sources. From January on the European Journal of Biochemistry will be named **"The FEBS Journal"** (see http://www.febs.org/Activities/Journals/febs_journals.htm)

## 5. Other FEBS Committees and working groups

FEBS is also aware of the rapid explosion of Science/Technology and of the repercussion these achievements are being reflected in Society and therefore have created a group of Science and Society to promote discussions on this issue. FEBS is also involved in other working groups such as Women in Science, Teaching Biochemistry and career of young scientists. FEBS has also created the Young Scientists Forum, which takes place every year immediately preceding the annual meeting.

## 6. FEBS interaction with other scientific organizations

FEBS interaction with other scientific organizations (EMBO, ELSO, and EMBL) has been fruitful for the science in Europe co-ordinating efforts, sharing experiences, and collaborating in areas of common interest. FEBS, together with the European Molecular Biology Organisation (EMBO; www.embo.org), the European Life Science Organisation (ELSO; www.elso.org), and the European Molecular Biology Laboratory (EMBL; www.embl.org) have taken the initiative in this respect by establishing, together with several other organisations, the European Life Sciences Forum (ELSF; www.elsf.org). The Forum was created to stimulate scientists to take a more active role in science policy issues. FEBS has been investing together with other scientific organizations in the founding of the European Research Council (ERC). This council will play an important role in the development of excellence in science and will help to set up young research groups.

## 7. Concluding remarks

In order to achieve all the aims of FEBS is required that the respective Societies are strong, well organized and with clear objectives. FEBS is well aware that it is only a small part of the European Scientific world and this also one of the reasons why FEBS likes to collaborate with other organizations. One of the possible future collaborations could be at the level of both undergraduate studies and of school teachers through the organization of courses which will promote their scientific knowledge. In order to have more information on the long range of activities visit the FEBS web site (www.febs.org).

*Science Education: Best Practices of Research Training for Students under 21*
*P. Csermely et al. (Eds.)*
*IOS Press, 2005*


# The Stockholm International Youth Science Seminar

**Hanna SJÖGREN**
*Förbundet Unga Forskare SIYSS,*
*Lilla Frescativ. 4c, 104 05 Stockholm*
*hannas@kth.se*

**Abstract.** The Stockholm International Youth Science Seminar was started as an initiative to increase attention and appreciation for the accomplishments of outstanding young scientists worldwide. Nearing its 30th anniversary in 2005, SIYSS has established itself as one of the foremost international youth scientific activity weeks, open to students of ages 18-26 years who have excelled within the realm of natural sciences. The participants, influenced intellectually and socially by this potent experience paralleling the Nobel Week, are readily inspired to serve as role models for younger talents they may come in contact with. Locally, Swedish students are encouraged by the seminar and its participants, to explore the vast opportunities available to them in science and technology while supporting organizations (governmental, academic, non-profit etc.) are made aware of the tremendous talents of young people across the globe. The SIYSS continuously strives toward enabling an even greater dissemination of inspiring influence on youth research training, talent recruitment, public awareness as well as cultural and multinational exchange.

## Introduction

Scores of scientific research ventures exist these days, serving as effective counterweights to the trend of increased polarization of societies, to a large extent targeting and isolating the scientific community. These projects, however, undoubtedly tend to concentrate in the societies of Western Europe, parts of Asia and Northern America where there historically has been a greater governmental and institutional emphasis on, and investment in, scientific efforts. In the age of globalization, not only is there a more apparent need to found new practices locally in countries where they do not exist, but also a call for for existing programs to open doors and set their enterprise on a more international arena. The SIYSS shares in the goals of the NATO-UNESCO Advanced Research Workshop to launch more of these research training practices world-wide, particularly in strained, divided societies of Central-Eastern Europe, Africa, the Middle-East and South America. Yet, the realization of attempts to find organizations to host youth initiatives in these developing countries may prove to be highly dependent on two characteristics of the target society; 1. the existence of advanced research projects and their willingness to devote time and effort to support youth recruitment, and 2. the existence of financial institutions willing to invest in this recruitment. Without positive feedback from these establishments, novel projects will likely find it utmost difficult to thrive.

With the birth of a new program, paying attention to the practical nature of the effort is of utmost importance. To send the message 'science is cool' and fully grasp the attention and motivation young people, it is essential to not only focus on the *contents*, but also the *staging* of these research training programs. Smart 'events marketing' and communication are key to establishing status for an event, a status that in turn strengthens the participants' identity and pride as researchers, and fosters a natural recruitment spiral.

The role of media, universities, government, private enterprise, and academic institutes functions to decrease alienation and promote integration with societal aspects of human life. Acknowledging these factors and clearly outlining a strategy to channel these groups toward the same goals and effects, stands as a fundamental building block for youth recruitment programs such as SIYSS. Our representation in the 2nd NATO-UNESCO Advanced Research Workshop serves to provide a "best-practice model" and comments in light of future global initiatives.

## 1. Background

In 1976 the Nobel Foundation, inspired by Science Service in the United States, sought to invest in youth active in science and technology by inviting seven students from three nations to participate in a week-long seminar parallel to the Nobel Week in December. The following year, the Swedish Federation of Young Scientists (FUF) was asked to co-organize the event, which proved to be a huge success. Consequently, the structure of the SIYSS organization shifted, and has since 1978 been entirely designed, planned and arranged on a voluntary basis by members of FUF, with support from a network of other organizations.

The core of the SIYSS program is clearly the momentary presence of a group of extremely talented young people together in Stockholm. What to do with this group, taking into account the vast possibilities, opportunities and potential available is far from an easy task.

## 2. Aims and goals

Adopting the overlying aim of FUF, SIYSS has sought to stimulate interest in science and technology in society as a whole with emphasis upon youth. In recent years, the structured goals of SIDA (the Swedish International Development Cooperation Agency) to contribute to international development cooperation have also to some extent been incorporated in the goals of SIYSS.

More explicitly, SIYSS has the ambition to directly or indirectly influence the following groups in the following ways:

| Youth (< age 18) | Increase personal interest for science and technology |
|---|---|
| | Mobilize activity and opportunity through elevated financial investment and socio-political efforts |
| Students (ages 18-26) | Increase students' motivation, influence and initiatives |
| | Raise awareness of and accessibility toward global research platforms through facilitated networking |
| Academic institutes | Highlight the importance of youth recruitment |
| | Decrease the communicative gap between youth and these institutes |
| | Give youth a better understanding of how the academic world acts and functions, of what resources are available |
| Governmental agencies | Delineate the role of talented scientifically active youth in societal development<br>Outline the role of governmental action in the success of scientific youth research-training recruitment and attention |
| Private enterprise | Make clear the necessity of private enterprise in research advancement and expansion, especially for youth during preliminary stages of projects |
| | Stress the value of research-trained youth in science and technology enterprise and the apparent influence of advanced research on the world market economy |
| Non-profit | Underline the tremendous, wide-spread effect that support for youth scientific research-training projects has in all societies |
| | Show how youth initiatives are exceptionally crucial to the advancement of underdeveloped and/or struggling societies. |
| | Make youth aware of the state in which societies other than their own are currently found in and how they personally and professionally can make a difference at home and abroad |
| Media | Amongst the general public, decrease scepticism/alienation and generate a more positive attitude toward science and technology on the whole |
| | Augment support for youth who wish to pursue a curiosity for science and technology, chiefly within their immediate social circle (family, friends, etc.) |
| | Heighten public attention for the most fascinating aspects and areas of contemporary scientific research |

Further, on the geographical plane, SIYSS adapts its goals and strategies for four principal focal groups; Sweden, the countries of our partner organizations, developing countries and the whole global arena. Getting a clear message across to all of these groups, and hopefully also generating some type of active reaction, is utmost difficult but far from impossible. Continuously reevaluating our rutines, certain things, however, just seem to work.

## 3. Voluntary organizations and outsourcing

Contrary to most other programs, the SIYSS Organizing Committee, like the Swedish Federation of Young Scientists, is in its entirety dependent on the time and effort of volunteers. The whole establishment is carried on year in and year out by the exclusively charitable endeavours of university students in Stockholm.

Though it may seem surprising to many, approaching planning of youth efforts in this manner has proven to be one of the most successful and effective elements of the SIYSS seminar. Many overlook the point that to inspire youth to take own initiatives, to inspire young people to want to make a difference, they must see that what is required in making a difference, and taking initiatives, is something they themselves are directly capable of, even if they start with nothing. If a young person is enthusiastic enough, and ready to invest time and effort to start a youth science competition, or seminar, or camp, they have with this energy come more than half way to realizing the event.

There are several examples of SIYSS participants starting competitions and/or organizations in their own countries, following their own personal experience of participating in the seminar. A national youth science exhibit in the Netherlands is a product of this, and a EU Contest-winner from Hungary was able to create a scientific research competition following his own SIYSS-experience in 2001.

In short, the voluntary nature of the organization and administration of the SIYSS has allowed the seminar to become a very personal and emotional event for both organizers and participants, which in turn breeds a stronger urge amongst them to speak of their experience and recruit other aspiring youths.

However, voluntary organizations have an inherent need to outsource many duties to other organizations or groups, seeing as ambitious voluntary efforts are still in no way as extensive as comparable professional commitments. Thus time is the apparent bottle-neck, but financing often creates problems as well. Moreover, these two factors don’t always go well together. The task of seeking direct financial aid and sponsors for a voluntary program is most often highly ineffective and time-consuming. Thus, SIYSS has adopted a strategy to work with organizations that can function as product sponsors or hosts for the variety of activities during the seminar week. Binding this type of support is much easier and in an elegant way molds large portions of the actual components of the SIYSS schedule.

## 4. The SIYSS program

The Stockholm International Youth Science Seminar begins each year on December 4$^{th}$ and ends on December 11$^{th}$. During this week, a intense program is designed to maximize social, professional and cultural exchange. Academic, governmental, non-governmental and professional types of organizations are involved in one way or another, strategically decreasing the organizational burden for the voluntary Organizers.

*Academia*

Visits at several scientific research centres at different universities in the Stockholm region give an in-depth account of what is currently happening in many areas of advanced research. For example, the Karolinska Institute hosts a luncheon and an afternoon series of workshops with researchers investigating areas related to the current Nobel Prize in Physiology and/or Medicine. Uppsala Universitet, Kungliga Tekniska Högskolan and Stockholms Universitet are also visited.

The Swedish Royal Academy of Sciences hosts an evening exclusively for the SIYSS participants probing the impact of the findings/inventions surrounding the present year's Nobel Prizes in Physics and Chemistry.

Being composed of university students, the SIYSS Organizing Committee has the advantage of several direct ties to student bodies in the region. These are easily involved in both social and academic aspects of the program.

*Governmental and non-governmental organizations*

The speaker or vice-speaker of the Swedish Parliament speaks privately with the group, outlining the governmental system in Sweden and guiding them through the Parliament buildings.

The Royal Court welcomes the SIYSS guests and usually makes a member of the Royal Family available for questions from the participants.

In the heart of Old Town, the students are guided by highly-qualified personnel around the Nobel Museum, a museum that allows for unparalleled insight into the history of both Swedish and international scientific research.

The Nobel Foundation most generously invites SIYSS to attend parts of the Nobel Week program, including the Nobel Prize-Awarding Ceremony and Banquet.

To promote better social and cultural exchange, several national, multinational and/or cultural interest groups are invited to host discussions, debates and other social activities open to SIYSS participants, students, researchers and the general public.

The Swedish non-profit organizations SIDA and Studiefrämjandet, in their cooperation with SIYSS, primarily focus on highlighting the state of third-world countries, what measures are being taken within the scientific field to aid with problems faced here, and what resources and opportunities young people have to impact this situation (these countries are unfortunately underrepresented among SIYSS participants and thus their situation not otherwise a natural part of the social exchange occurring during the seminar).

The Swedish International Water Institute (SIWI) is engaged during the week in informational work about global water research.

*Other*

Fairly simple measures are taken to guarantee social exchange outside the "formal program" schedule.

By accommodating participants at a youth hostel, as students share rooms and facilities they find it easier to communicate and bond with each other, which also minimizes isolation and alienation of any single participant.

Each evening a dinner with a specific "theme" is arranged. Traditionally this may include a get-together (promotes participants to get to know each other on a more personal level), an international dinner (an evening filled with show-and-tell about the participants cultural heritage) and an etiquette dinner (informal and exaggerated practice in how to behave at the Nobel Banquet).

## Concluding remarks

The enduring tradition of the Stockholm International Youth Science Seminar proves that elaborate and influential endeavours geared at youth can thrive even in the absence of any greater means of funding or professional support. Voluntary organisations and the energetic individuals that propel them have awesome potential simply waiting to be realized. In countries where facilities and resources are and have been limited, such as large portions of Eastern Europe, the Middle East, South America and Africa, modelling efforts after the SIYSS example may prove highly successful. Organizations seeking assistance or support for this type of activity are more than welcome to contact the SIYSS Organizing Committee for guidance. Organizations interested in sending a participant to SIYSS, especially from countries in the aforementioned regions, are kindly asked to contact us.

For more information: *www.fuf.org/siyss* or send us an e-mail at *siyss@fuf.org*.

**References:**

[1] The Stockholm International Youth Science Seminar: http://www.fuf.org/siyss
[2] The Swedish Federation of Young Scientists: http://www.fuf.org
[3] The Nobel Foundation and The Nobel Museum: http://nobelprize.org
[4] The Swedish International Developmental Agency: http://www.sida.org
[5] The Swedish International Water Institute: http://www.siwi.se
[6] The Royal Swedish Academy of Sciences: http://www.kva.se
[7] The Swedish Academy: http://www.svenskaakademien.se
[8] The Royal Court of Sweden: http://www.royalcourt.se
[9] The Swedish Parliament: http://www.riksdagen.se
[10] The Karolinska Institute: http://www.ki.se
[11] The Royal Institute of Technology (KTH): http://www.kth.se
[12] NATO Security through Science: http://www.nato.int/science/
[13] UNESCO Science programmes: http://www.unesco.org/science

*Science Education: Best Practices of Research Training for Students under 21*
*P. Csermely et al. (Eds.)*
*IOS Press, 2005*


# Use of Mentoring in Precollege Science Education Programs of Howard Hughes Medical Institute Grantees

**Jill CONLEY**
*Howard Hughes Medical Institute*
*4000 Jones Bridge Road, Chevy Chase, MD 20815 USA*
*conleyj@hhmi.org*

**Abstract.** The Howard Hughes Medical Institute supports precollege science education programs that create dynamic partnerships involving scientists, educators, students, their families, and their communities. These partnerships engage and excite children by presenting them with challenging inquiry-based learning experiences in science. Within many of these programs, mentorship provides an excellent means to provide equity in access to effective educational products and practices and inspires young people traditionally underrepresented in scientific professions to pursue scientific activities and careers.

## 1. HHMI Support of Precollege Science Education

The Howard Hughes Medical Institute (HHMI) is a medical research organization dedicated to supporting outstanding basic biomedical research as well as science education at all levels.

HHMI's precollege science education program funds educational and research institutions that target children from preschool through high school along with their teachers, their families, and their communities. By creating partnerships among scientists, educators, and the public, HHMI fosters increased public interest in and understanding of science with the goal of improving science literacy. Scientists provide contemporary and accurate content for teachers, who in turn refine and present more engaging curricular materials to the students. Parents and community groups are active participants in the children's education, reinforcing the fact that learning doesn't end with the close of the last class period.

HHMI promotes science education reform. Students need to be active participants in the learning process to become engaged in and excited by science. Program grantees provide authentic hands-on research experiences to students and teachers, avoiding standard didactic classroom practices and introducing participants to the thrill of discovery. Immersion into research sharpens the critical thinking skills of the novice investigators, challenging them to question, hypothesize, and experiment; answers lead to further questions, and the student becomes enmeshed in the wonder of stretching his or her own personal understanding. This independence from otherwise rote, structured acquisition of information makes the student the architect of his / her learning process, empowering him / her to pursue information creatively in a meaningful and productive way. Once captivated by the process of inquiry, students see that science is not an isolated endeavor but is instead an integral part of everything they encounter.

An important aspect of education reform is to ensure that girls and traditionally underserved populations (or populations traditionally underrepresented in the sciences) have equity in their access to new science education content and practices. Not only is this critical to ensure a diverse and highly qualified pool of scientists for the next generation, but it is a vital element in producing a society with the critical thinking skills necessary to address the increasingly technologically sophisticated issues faced by voters and decision makers who need to make intelligent, informed choices.

Successful programs designed to reform science education by facilitating participation of underserved audiences rely upon effective partnerships among stakeholders. One set of partnerships that provides a valuable tool to recruit underserved populations into science involves mentoring. Provision of role models and mentors introduces young people to the idea that a scientific career can be an attainable and attractive goal, and can inspire children to overcome perceptual challenges to realize their potential. Mentors can be drawn from myriad sources – scientific researchers and educators, professional groups, school (both teachers and older students), and the community. HHMI supports a variety of programs that use mentors to engage children, as the following examples illustrate.

## 2. HHMI Programs in Precollege Science Education

Currently, HHMI follows three different approaches to improve precollege science education in the United States.

- One initiative supports institutions of informal science education, e.g., science museums, zoos, aquaria, and botanical gardens, to form partnerships with schools and community groups. These institutions use unique mechanisms to deliver their science content and attract a broad target audience that voluntarily seeks out a science education experience. They employ educators known for their skill in engaging a diverse public, and who generally communicate to family groups with varied interests rather than to students in a formal classroom setting.

- Another initiative supports medical (and dental and veterinary) schools, academic health centers, and biomedical research institutions, who have powerful science resources that can be used to enrich science education by providing students and teachers with laboratory research experiences. They employ researchers who generate the state-of-the-art scientific content that is often difficult to disseminate to the precollege classroom, and actively involve these researchers in precollege science education.
- The third initiative involves direct partnerships between HHMI and two local public school systems. These partnerships are designed to stimulate the interest of students in science and science-related careers by using locally available science resources to enhance education for students of all ages as well as to promote teacher professional development.

The programs selected as examples for this discussion derive from the first two initiatives, since a presentation from the third initiative is included separately within this collection.

### 2.1. The Ecotarium

The Ecotarium, in Worcester, Massachusetts, created partnerships with the local housing authority and three high schools for its year-round work-based program for inner city high school students. This program provides students, primarily from low-income and minority communities, with opportunities to learn environmental science and receive job training. Participants progress through different levels of instruction, gaining responsibility and new opportunities. As they achieve mastery of content in areas such as forest ecology, recycling, and freshwater ecosystems, they serve as mentors, training younger students. Eventually, some of the mentors become interns at the museum. Once the students have mastered the material, they deliver programs to the public attending the museum.

Museum staff working with the students faces enormous challenges. They provide tutoring to incoming students whose academic background is incomplete, and have been truly gratified to experience the burgeoning interest of these students in science. Student participants reported enjoying school science classes. Many students saw themselves as real "learners" for the first time.

The mentoring program helps students navigate the graduation process and college and employment decisions, and includes visits to colleges in the region to encourage students to start thinking about the importance of college when choosing a career. Although most of the high school students had never considered education beyond high school when they began the program, all of the high school seniors in the most recent cohort will attend college in the fall.

### 2.2. The Robert C. Byrd Health Sciences Center of West Virginia University

The Health Sciences and Technology Academy (HSTA) at the Robert C. Byrd Health Sciences Center of West Virginia University in Morgantown reaches out to disadvantaged high school students to encourage them to consider careers in math, science, and health. Talented students who might be economically (or otherwise) hindered from pursuing such professions are nurtured through a series of mentored activities both during the school year and summer vacation. In addition to increasing the rate of college attendance of disadvantaged students from the region, the larger goals of the program are to improve science and math education, to help communities develop leadership of their youth, and to increase the number of health care providers in West Virginia's under-served rural communities.

HSTA is a partnership among the university, the West Virginia Health Education Partnership, and many local communities. Participating students are nurtured through after school and summer activities as well as through involvement in local science clubs that include mentors who are teachers, health professionals, and scientists in the community. Students and their teachers visit the university campus in the summer for fun hands-on activities that include clinical, laboratory, and classroom training, as well as enrichment activities integrating math and science curricula with leadership development, self empowerment, communication skills, study skills, and multicultural sensitivity. During the school year, the program supports community-based science projects that are mentored by teachers, health profession students, and volunteer community leaders.

HSTA tracks the accomplishments of participants in college and professional school, and has been so successful that the West Virginia state legislature now awards all program graduates four years of free tuition at any public college or university in West Virginia. The program currently serves 22 of West Virginia's 55 counties.

### 2.3. Fox Chase Cancer Center

The Fox Chase Cancer Center in Philadelphia, Pennsylvania, designed the Partnership for Cancer Research Education to help middle and high school students understand science better by participating in cancer research. The program includes:

- A research program aimed at providing entire classrooms of middle and high school science students with an active research experience. Teachers work with scientists to design open-ended research projects in which the students do experiments, analyze data, and draw conclusions over the course of the school year. The scientists visit the classrooms, the students visit the labs, and both communicate via email. At the end of the school year, the students present their work to the community.

- An intensive mentored research experience for selected students who have demonstrated interest and ability in science. Students present their work to their peers and at a poster session at Fox Chase. Many participants use their research as a basis for science fair competitions, and some have published their work in peer-reviewed journals.

- A mentored teacher professional development program in which teachers conduct research in their scientist-mentor's lab during the summer.

Since 1999, 49% of the labs of Fox Chase scientists have participated in this program, and 59% of the students have continued doing research with their mentors after the one-year program.

**2.4. Yale University School of Medicine**

Yale University is part of an urban Academic Health Center, and is "committed to providing learning environments for young people in New Haven by which human potential can be fully realized." Through this project, Yale wants to improve students' science knowledge and content skills by developing their critical thinking and problem solving abilities. They use mentoring to motivate the students, and develop new curricular resources.

The program, called the Science Collaborative Hands-On Learning and Research (SCHOLAR) Program, targets high school students and provides a problem-based learning (PBL) approach that incorporates an integrated science curriculum, research and medical internships, and mentoring. SCHOLAR includes a tuition-free residential summer science program for students enrolled at a partner urban high school, and is seen as a model for school – university collaborations. Program faculty includes partner high school science teachers, Yale faculty, and Yale professional, graduate, and undergraduate students.

The student participants, parents, the university, and the high school teachers and administrators are very enthusiastic about the program. The interactions between the students and their mentors, and the relationships formed, appear to have a lasting positive effect. Approximately 80% of SCHOLAR student participants return after their initial summer experience to continue for an additional summer (or more).

With respect to the success of the SCHOLAR program, Yale reported that:

- Not only would every graduate of the program be heading to a four-year college in the fall, but a significant portion (40%) of the students were in fact heading to college as their family's first representative in higher education.

- The average SCHOLAR graduate gained admission to 4.4 colleges, and no graduate had fewer than three admissions. In addition, 88% of the SCHOLAR graduates were not rejected from a single college to which they applied.

- 90% of the SCHOLAR graduates reported that they planned on majoring in some sort of scientific field in college (as compared to 14% of minority students nationwide). Career ambitions were also oriented along the scientific lines and were quite high: among those who specified, careers that involved the acquisition of an advanced degree were named as future job goals for 71% of the students. And although the U.S. science and engineering workforce has only 7% minority employees (Black, Hispanic and Native American), 90% of SCHOLAR participants stated they sought employment in the scientific realm after graduation.

### 2.5. The Imaginarium

The Imaginarium, in Anchorage, Alaska hopes to stimulate an interest in science in students in remote rural Alaskan villages, to engage their families and communities in the excitement of inquiry-based learning, and enhance curricular materials for Alaskan teachers. The Imaginarium has formed partnerships within the scientific, cultural, and educational communities to develop programs that have accurate content and that are culturally appropriate.

The population served by the Imaginarium is an exceptionally difficult one for which to affect change; the communities are physically isolated, extremely impoverished, and distrustful of "outsiders." Because of the extreme climate and geographic conditions in rural Alaska, access to villages is often limited primarily to air transportation. There are no local universities or large centers of learning. Thus, there are few extracurricular educational opportunities or role models, and little encouragement for students to engage in scientific inquiry let alone pursue a career in science. Few aspire to leave the village, and most are unaware of opportunities available to talented students in science and math.

The program uses a Science Outreach Caravan to deliver activities that include inquiry-based school assembly programs and hands-on classroom presentations covering the earth, life, and physical sciences. These programs stress observation, conceptualization, hypothesizing, and testing, using a participatory and hands-on approach. Teachers are mentored in workshops to enhance inquiry-based teaching skills and introduce participants to hands-on, interactive classroom science lessons and activities. Community science festivals are included to provide a social event for the villages, drawing families into the excitement of science.

The Imaginarium staff realized the importance of gaining community acceptance for their program, so they worked with local tribal elders to identify needs and appropriate content. They have learned that the best source of native knowledge lies with the local educators and the community, and that without their involvement in translating that knowledge into relevant and appropriate curricular materials the program would be met with resentment rather than enthusiastic engagement. However, despite these inherent challenges, the Imaginarium has doubled the total number of general science programs delivered within Alaska. They report receiving "unprecedented positive responses from teachers, students, and community members. This positive response reflects the relevance of the programs offered and an increasing openness to science education, thereby producing students more likely to pursue science."

### 2.6. Columbia University College of Physicians and Surgeons

Columbia University in New York City aims to enhance the interest and performance of underserved New York City high school students in science by mentoring their teachers in laboratory research. They believe that teachers involved in the day-to-day conduct of science will be better able to motivate their students in science. Columbia University faculty host teachers for an annual eight-week summer laboratory research experience for two years, and these teachers work closely with their scientist mentors. Once back at their schools, teachers share their experiences with their students and fellow teachers, and prepare new lesson plans to communicate what they have learned. Staff meets weekly with teachers to help them connect their research experience to the classroom, forming a professional learning community. Discussion topics include educational technologies, pedagogy, and content underlying their research projects.

Columbia has made excellent inroads into evaluating the success of their intervention. They assess the impact their program has had on students of participating teachers by looking at quantitative measures and comparing these results with those of students of teachers who have not participated in the program. They have examined indicators such as student interest (measured by participation in science clubs and science competitions) and student achievement (measured by test scores). They report that "in the academic year prior to entry of teachers into this Program, ~32% of their students passed a New York State Regents science exam compared to ~37% of students of other teachers in the same schools and science departments. In the two years following their completion of this Program, ~43 and ~48% of their students passed the Regents science exam, compared to ~32 and ~37%, respectively, of students in science classes of other teachers in the same school."

## 3. Conclusions

More information on these programs and others funded by HHMI, along with contact information for each program director, may be found on the HHMI website at www.hhmi.org/grants/office/precollege. These programs are representative of some of the innovative, practical, and effective efforts aimed at drawing students from all communities into the excitement of science and widening the scientific pipeline to become more inclusive.

In each case, partnerships integrate the various knowledge and skill sets required to design effective educational materials and facilitate the incorporation of the ideas into practice. Mentors provide inspiration, support, and experience, and may be students, teachers, scientists, or respected community members. Likewise, the benefits of mentorship extend beyond a direct relationship with a student to targeting teachers; one teacher, in turn, can share her / his enthusiasm for the experience of discovery to many students. Role models illuminate possibilities, creating new opportunities for students previously bound by unnecessary limitations.

Scientists, educators, families, and other community members play an important role in engaging children – all children – in science education and in preparing them for the technological complexities of the society of which they are a part.

*Science Education: Best Practices of Research Training for Students under 21*
*P. Csermely et al. (Eds.)*
*IOS Press, 2005*


# The Role of the Inner and Outer Environment on Science Education and Research Training

**Otto JOKLIK**
*EUREKA International,*
*Gersthoferstrasse 120, A-1180 Vienna, Austria*

**Abstract**. Both inner and outer environments are crucial in the development of talented students. Since many elements of the outer environment are given, we should take a very good care for the development of the inner environment of the student. In this contribution brief guidelines are given to achieve this goal with the additional aim to draw students' attention towards the application of their basic or applied science ideas.

The inner and outer environment has a substantial influence on science education and research training. The outer environment comprises a clean air, safe drinking water and a healthy housing. It is evident that little can be done to influence these conditions. They are within the responsibility of the public sector.

Under the inner environment we understand the life within a healthy, well functioning family and their full support attending the elementary and secondary school. Children should play with constructive three-dimensional toys to foster their creativity and imagination rather than to spend its time before a TV screen. The further school education of the growing youth should be performed preferably in gymnasium, where the education to principal thinking and also understanding of complex processes is ensured. The young student should be encouraged to be a generalist with universal interdisciplinary interest than an over-specialist with a narrow view of a singular field of interest. The student should be given the opportunity to work in individual small groups in practical experiments as his or her creativity will continuously grow with this experience. The student's inner environment should foster his or her fantasy, which is the basis and drive in the future research work and scientific performance. The fantasy and imagination plays a vital role in science. They foster the creativity finally to a successful research and scientific achievements.

The efforts of advanced students should be directed towards applied research. They should concentrate on innovations and inventions with the objective to improve and develop existing technologies in the fields of machinery, equipment and industrial production plants. In such a way the Quality of life could be substantially improved, which is of utmost importance in the developing countries. The improvement of existing machinery, equipment should start by a thorough study and implementation of the best of existing technology. A virtual technology development should follow, later on assisted by CAD (Computer Aided Design). The construction of three dimensional models should be followed by prototypes and pilot plants, which already allow an assessment of a feasibility of a project.

There is a long way from the initial idea to a commercial product. Many collateral problems arise during the developments of project. The advanced students should be confronted with them already at an early stage of development. The most important ones are:

1. Intellectual property, patents, trade marks
2. Negotiations for sales of technology and know-how, licensing, contractual aspects and conditions
3. Publications, marketing, Internet
4. Environmental aspects of the new technology
5. Market development control, control of competitors
6. Financing, joint—ventures, risk capital
7. Project management and control

An experienced project management is essential for the implementation of a new technology. An invention or innovation occurs in the field of fundamental sciences only exceptionally. In most cases the innovation or invention will be in the field of applied sciences. Here the evaluation criteria of a project plan may be formulated as follows:

**Table 1.** Evaluation and judgment criteria.

| Creativity | 30 points |
|---|---|
| Technology | 20 |
| Social importance | 20 |
| Commercially | 20 |
| Environment protection | 10 |
| | 100 points |

According to the number of points attributed the invention or innovation can be classified as follows. These criteria may serve as a guideline for the further activities of an inventor or innovator.

**Table 2.** Evaluation thresholds.

| 90/100 points | Excellent |
|---|---|
| 80/100 | Very good |
| 70/100 | Good |
| 60/100 | Acceptable |
| 50/100 | Not recommendable for follow—up efforts |

Evaluation is not only an endpoint of a student's project. A continuous and personal feedback is highly important for maintaining the motivation and ensuring the continuity of creative ideas of our talented students. This is important point, where the outer environment is affecting the inner environment: the innovative process of the student.

# Session III

# Successful Practices of Research Training – Israel, the USA and Korea

*Science Education: Best Practices of Research Training for Students under 21*
*P. Csermely et al. (Eds.)*
*IOS Press, 2005*



# Science Education for Gifted Students in Israel Via Distance Learning

**Shlomit RACHMEL, Rachel ZORMAN, Taly BEN–YEHUDA, Zeev STOSSEL**
*Department of Gifted Students, The Ministry of Education,*
*2 Dvora Ha-Nevia St. Jerusalem, Israel*
*shlomitra@education.gov.il*

**Abstract.** The Department of Gifted Students in the Ministry of Education in Israel aims to enable gifted students in different regions of the county to engage in high quality studies in the sciences. One of the most notable efforts of the department to achieve this goal is promoting distance learning of science. In this paper, we will explore the nature of distance learning, its advantages and limitations and whether it is appropriate for gifted students. We will describe four distance learning courses in the sciences that were piloted in the virtual school established by the Department of Gifted Students in the years 2002-2004 and present results of evaluation of these courses. Finally, we will discuss the implications of distance learning for the education of gifted students in the future.

## Introduction

One of the major goals of the Department of Gifted Students in The Ministry of Education in Israel is to provide accessibility to innovative, quality science curricula for gifted students in all parts of the country, including areas in the periphery. In line with this goal, the Department operates a virtual school on an internet website that enables students to engage in distance learning via virtual science courses. In this paper, we will delineate the nature of distance learning, detailing the conditions for successful distance learning. We will proceed to describe the goals and design of four distance learning courses for gifted junior high school students. The results of piloting these courses during a two year period from 2002 to 2004 will be presented. Furthermore, the implications of this type of learning for future online programs for the gifted will be discussed.

## 1. The Nature of Distance Learning

Distance learning involves a system of interaction between learners and content matter on various websites on the internet. Distance learning may be asynchronous or synchronous, or may combine the two modes of learning [1]. Asynchronous learning is based on content and resources stored on web servers. Synchronous learning involves real-time interaction on the Internet between a teacher in one location and students in various locations through such means as video conference or electronic chats.

The learning environment on the internet attempts to provide gifted students with the vital conditions they need for effective learning, knowledge construction and creativity, in order to help them realize their potential for their own benefit, as well as for the benefit of society. Distance learning creates and facilitates a challenging learning environment characterized by flexibility and dynamism in addressing complex learning tasks and enhancing creativity; requiring initiative, assumption of responsibility for one's own learning, and reflection about thought processes.

### Advantages and Disadvantage of Virtual Learning

In designing distance learning courses, one has to take into consideration the advantages and disadvantages of the internet medium. Based on our experience with distance learning and on research in this field [1], the advantages of such learning include:

- No time limitation in asynchronous learning, where students interact mainly with contents and resources on the web at any time that is convenient to them. This enables students to utilize their time more effectively.
- No location limitations, since distance learning can happen anywhere there is access to the internet. This is especially important to students in the periphery, who do not often have access to special science courses and to unique teachers in their vicinity.
- Access to the most updated information resources available, since learning relies on websites, not on textbooks.
- The possibility of making dynamic changes in course content and tasks as a result of the interaction between the student, the instructor and the website content.
- Comprehensive use of multimedia, with the possibility of utilizing various visual aids and simulations of phenomena to enhance learning.
- Student paced learning, since students learn according to their own rate, skimming over topics that they already know and devoting more time to learn subject matter that is new to them.
- Documentation of all lessons, which enables students to revisit the lessons whenever they want to and require the teacher to plan the lessons more carefully, hence improving the quality of instruction.

However, distance learning also has its disadvantages. These include:

- Limited personal interaction with an instructor. Students do not see their teacher and vice versa. The lack of immediate feedback between teacher and students prevents the teacher from adjusting the content and instruction in real time.
- Lack of direct personal contact between the students, when there is limited social contact and exchange of ideas between students.
- The need for technical support for users to enable them to use all the resources on the web.
- The need for a significant time investment to investigate various websites.
- The need for considerable self discipline to stay on task.
- The fact that success in virtual learning depends, in part, on the basic computer skills of the students.
- The lack of direct instructional media, such as actual labs and experiments and real time discussion groups.

The above problems may lead to a considerable drop out rate from distance learning courses. Salant [1] notes that various follow-ups of students in Israel and abroad report a 25-50% dropout rate from distance learning courses. However, a study by Dallas and Franklin [2] showed an increased interest in virtual learning courses in comparison to traditional courses in high school. Moreover, evaluations of achievement in high school conducted by Fleming and his colleagues [3] showed greater equity of access to learning opportunities for all students in distance learning courses, as well as enhanced achievement in distance learning courses, especially in foreign languages and science.

## Capabilities and Skills Required for Distance Learning

Distance learning requires certain capabilities and skills above and beyond the skills needed to engage in face-to-face learning.. The most important capability is self discipline. Since there are no defined time and location frameworks for class meetings in asynchronous distance learning, students have to be self disciplined. In addition, they have to possess internal motivation to learn, since the teacher's ability to fine tune the courses to motivate students is limited. Moreover, students must be independent learners. They have to cope with problems that come up during the learning process on their own, without the help of the teacher or their peers. Further, they have to formulate questions that can help them obtain information and skills to cope with their difficulties. Last, but not least, distance learning requires considerable technical skills to use the internet as a learning tool

## Gifted Students and Distance Learning

Gifted students have unique characteristics that result in special needs. Distance learning may address these needs quite effectively in the following ways:

- Gifted students are effective consumers of knowledge. Distance learning enables students to link to various websites that present opportunities to enhance their learning in breadth and depth.
- Gifted students often learn in a very fast pace. With its individually paced learning, distance learning enables students to study at their own pace.
- Gifted students often exhibit a high ability to study independently. Since independent learning forms an important basis for distance learning, the chances of gifted students to succeed in distance learning are higher than the chances of their non-gifted peers.
- Gifted students are very curious and internally motivated to learn about topics of interest. This is important for successful distance learning, since the teacher's influence on academic motivation is considerably lower than in face-to-face learning.

The Department of Gifted Students in the Ministry of Education in Israel is committed to create various educational frameworks to address the needs of gifted students: To provide every gifted student a work environment suited to his/her needs and adapted to his learning style, such that s/he will be able to develop his/her capabilities to the greatest possible degree. In this environment, the teacher serves as a mentor, advising, supporting and directing the student in the educated and controlled use of diversified sources of information; among them, the use of technological tools. The nature of student work in distance learning encourages divergent thinking, multi-disciplinary thinking, development of creativity and originality in the problem solving process. Thus, the teacher must be capable of cultivating and guiding these processes, and helping to design applications of understanding at the highest level.

At present, gifted students are usually offered enrichment once a week in magnet centers, in extracurricular afternoon clubs or in year round special classes and in two special schools. However, some gifted students do not enroll in these educational frameworks because they live far away, usually in communities in the periphery. In these communities, the additional enrichment opportunities provided under the auspices of the Department are rather limited due to the distance from institutions of higher learning; and, in many cases, because of difficult economic conditions. Hence, there is a gap in learning opportunities offered to gifted students in large urban areas as opposed to those living in small, remote communities.

In many cases, gifted students feel isolated in the regular system, or they may feel that they have too much of an academic burden, especially in secondary school. Distance learning enables these students to take high quality courses on topics of interest when it is convenient for them. This enables them to study in a virtual environment together with students who have similar interests and abilities. Furthermore, distance learning leads to higher quality work than in heterogeneous groups where gifted students are few and far between. It enables gifted students to acquire highly important learning skills that are valued in business organizations and in universities.

## Conditions for Successful Distance Learning

How can we capitalize on the advantages of distance learning, while coping with its disadvantages, in order to enable successful distance learning? Investigators who designed and administered such courses in Israel [1] and abroad [4, 5, 6, 7, 8] propose several essential components that ensure success distance learning. These components relate to technical support, and, more importantly, to various modes of interactive support for the students.

Technical support includes such aspects as:

- Enhancing computer and internet skills before starting the courses.
- Providing technical support for problems arising during the courses.
- Creating parallel channels of communication, such as attached word files and links to various internet sites.

Various modes of interactive support for the students include:

- Providing continuous feedback to students by the instructors, utilizing e-mail and internet chat forums.
- Presenting content in hierarchical, concise and clear manner to prevent information overload.
- Designing structured, organized activities with clear expectations for performance.
- Creating a mutual contract for learning, detailing the responsibilities of students and instructors, thus promoting commitment to the course.
- Creating a virtual community of learners via e-mail, chat forums and computer conferencing, encouraging students to pose questions, discuss various issues with the instructor and with their peers and promote group projects on topics of interest.

## 2. Goals and Objectives of Distance Learning Courses in the Sciences

The goals of the distance learning courses developed for the Department of Gifted Students are threefold:

1. Presenting an intellectual challenge that encourages students to investigate several subject areas in greater depth, and requires personal commitment and responsibility for learning.
2. Providing an opportunity for students to experience distance learning as a basis for future learning.

3. Creating conditions that will turn the learning into an enjoyable experience and a process that continues for a lifetime, in light of the fact that knowledge doubles itself within a short time.

   The specific objectives of the courses include:

- Creating an opportunity for students to utilize different learning styles, especially independent learning.
- Providing an opportunity for intellectual meeting of students with similar interests.
- Developing an interdisciplinary perspective of science.
- Providing a unique learning experience with content experts from the university, who expose students to different methods of scientific investigation and analysis.

## 3. The Design of the Virtual School

During the first eight years of operation of the internet site, we focused on scientific surveys and complex units in a broad variety of fields of knowledge which were written by academic experts and suited to gifted students, grades 7 and up. Subjects such as cellular telephones, the problem of the rain forests in Brazil, or a literary analysis of the Harry Potter series had a tremendous number of hits. Each survey was accompanied by questions for thought, links to additional sites dealing with the same subject, referral to relevant illustrations and animations. Likewise, the writer of the series facilitated discussion groups and answered questions for a period of two months from the date that the survey was placed on the site. Every month, we put a new survey on the site.

Despite the large number of hits, we realized that some of them were by students from the general system seeking a good source of information in Hebrew, and discovered that these surveys met their needs. Likewise, the contact between the writer of the material and the student was non-committal and random. Students were responsible to perform the tasks on their own and they got no feedback, except in those cases where they initiated contact with the writer of the material.

In light of the above limitations, we made a decision in 2002 to change the concept of the internet site. It was clear that in order to meet the needs of the gifted students, especially in the periphery, the activities had to be organized systematically and a virtual school had to be established. Four courses were prepared by experts from academia and from the Department. We approached seventh to ninth grade students in the special centers and there was a great show of enthusiasm. The choice of relatively older students was based on professional considerations. These students usually master the necessary technological skills, know English well and are ready for independent learning. We limited the number of students in each course to 30, in order to facilitate feedback and formation of a virtual community.

Our virtual courses were piloted over a two year period in the years 2002–2004 with Tel-Aviv University and the Center for Educational Technology. The topics of the courses included the senses, artificial intelligence, the history of mathematics, and environmental ethics. The modes of learning employed in these courses develop divergent thinking, and are based on capabilities of higher order thinking, and use of diverse information sources, which characterize gifted students. Likewise, ethical aspects and interdisciplinary connections are addressed.

Each course introduced various scientific concepts and examined them from philosophical, psychological and social perspectives. For instance, in the course on the senses, students studied the physiology of the senses and addressed relevant philosophical questions, such as: Is there an objective reality? Do different people perceive reality in the same manner? In the course on artificial intelligence, students studied the physiological and psychological bases for human thinking and the technologies that attempt to mimic thinking. They dealt with philosophical questions such as: Will the computer be able to think like us some day? Do we want the computer to think like a human being? What does that mean in relation to the issue of body and soul? In the course on the history of mathematics, students learned about the development of mathematical thinking through the ages and attempted to solve mathematical problems that mathematicians labored on in the past, using the knowledge that was available to them at that period. In the course on environmental ethics, students studied ethical issues related to the environment. They learned about ecological systems and how modern life created environmental problems. They discussed several philosophical approaches to the interaction of humans with their environment and analyzed actual issues (such as the question of whether it is appropriate to work to limit population growth in order to deal with environmental problems) according to these approaches. Furthermore, they engaged in a simulation game attempting to determine global policy to deal with the global warming effect.

The courses were comprised of twelve weekly units that were presented on the internet. Each unit included an introduction, textual material on the topic, links to various information resources and student performance tasks, which were required to be completed each week. The instructors provided students with feedback on the tasks that they turned in and on their participation in various activities. A personal meeting of instructors and students was organized once every course in Tel-Aviv University. Students spent a whole day in the university. They met with their instructors. They visited a facility that was relevant to their studies. For instance, students participating in the course on artificial intelligence visited a robotics lab, whereas students participating in the course on environmental ethics visited the zoological garden in the university. At the end of the day, students participated in an academic lecture that introduced them to various enrichment topics.

As a result of feedback from students and instructors at the end of the first year, some content adjustments were made, but more importantly, certain features were added to the interaction with the students to enhance the support provided to them. Each week, the instructors provided two office hours on the internet for the students. The instructors initiated interaction with students who were late in completing assignments, offering extra help when needed. Student chat forums were organized, as well. In addition to online feedback to students on each assignment, students and parents received two progress reports by mail, after four weeks, and after ten weeks, as well as a report card at the end of the course.

In the first year of operation, the target population consisted of gifted junior high school students in self contained classes that participated in the courses as part of their daily schedule at school. In the second year of operation, the target population included all interested gifted junior high school students nationwide, who were willing to devote time and energy to engage in distance learning. Thus, in the second year, participation was voluntary and based on personal interest.

## 4. Results of Piloting the Distance Learning Courses

The results of piloting the distance courses over two years are encouraging. In the first year of operation, 50% of the 115 participants completed the virtual courses. In the second year of operation, the total number of participants increased to 155. Moreover, the percentage of students completing the courses rose from 71% in the first semester to 88% in the second semester. This significant change is probably due to voluntary student participation in these courses, to some changes in course content and to the enhanced support provided to students that were detailed above.

Student feedback in the second year of operation is very positive. In an attempt to investigate the sources of motivation for engaging in virtual learning in comparison to learning in their school program, students were asked to rate the following sources of motivation for learning on a scale ranging from 1, denoting very low degree of agreement, to 5, denoting very high degree of agreement. Means and standard deviations (SD's) of their ratings, as well as t-scores indicating whether the difference between virtual learning and school learning is significant, are presented in Table 1.

**Table 1.** Student Report Concerning the Sources of Motivation for Learning (Means, SD's, and T-Scores).

| | **Degree of motivation to learn in a virtual course** | | **Degree of motivation to learn in a school course** | | |
|---|---|---|---|---|---|
| | **Mean** | **SD** | **Mean** | **SD** | **T – Score** |
| **I want to get High grades** | 3.78 | 1.94 | 4.29 | 1.04 | -1.81 |
| **I enjoy studying** | 4.40 | 0.87 | 3.23 | 1.21 | ** 5.22 |
| **I am interested & curious about the subject** | 4.77 | 0.66 | 3.37 | 1.31 | ** 6.06 |
| **I would like my parents to appreciate me** | 2.42 | 1.39 | 2.61 | 1.58 | -1.38 |
| **I would like my teachers to appreciate me** | 2.15 | 1.52 | 2.82 | 1.45 | **-3.09 |
| **I would like my friends to appreciate me** | 1.64 | 1.13 | 2.13 | 1.42 | **-2.98 |

*p <.05 **p<.01

This table demonstrates that interest, curiosity and enjoyment serve a more significant motivating role for distance learning than for school learning, whereas teacher and peer appreciation serve a more significant motivating role for school learning.

Students were asked to rate the most salient features of the courses, such as the contact with the instructor and the nature of the learning in the course. Their feedback is presented in table 2 in means and standard deviations (SD's) on a scale ranging from 1, denoting very low degree of agreement, to 5, denoting very high degree of agreement.

**Table 2.** Student Feedback on Features of Distance Learning Courses (Means and SD's).

| Course Features | Means | SD's |
|---|---|---|
| The feedback on assignments was clear | 4.51 | 0.84 |
| The face to face meeting with the instructor was helpful | 3.02 | 1.31 |
| I had continuous contact with the instructor | 3.49 | 1.17 |
| I took extra care to complete my weekly assignments | 4.38 | 0.93 |
| The atmosphere in the course was open and encouraging learning | 4.3 | 0.82 |
| I felt comfortable contacting my instructor with questions & problems | 4.37 | 0.84 |
| The chat forums in the course helped understand the assignments | 3.18 | 1.01 |
| Learning in the course demanded more in depth thought than in school | 4.54 | 0.73 |
| Learning in the course was challenging | 4.42 | 0.64 |

From the above feedback, it is clear that the interactive support that was provided to students in the form of instructor feedback on assignments, questions and problems that arose, and the nature of learning in the course, involving challenges and in depth thinking, created an open atmosphere, enhancing learning to a great degree. Furthermore, student satisfaction from various features of the course was quite high, as table 3 demonstrates.

**Table 3.** Student Satisfaction from Various Features of Virtual Courses (Means and SD's).

| Course Features | Means | SD's |
|---|---|---|
| The chat forums were interesting | 3.88 | 1.02 |
| Learning in the course was interesting | 4.58 | 0.5 |
| Social contacts were formed with students from other schools | 2.44 | 1.38 |
| I enjoyed the assignments in the course | 3.93 | 0.94 |
| I feel that I learned the topics in the course in depth | 4.24 | 0.9 |

Student feedback, presented in table 3 in means and standard deviations (SD's) on a scale ranging from 1, denoting very low degree of agreement, to 5, denoting very high degree of agreement shows that most students are very satisfied with the courses. They expressed special satisfaction from the in depth studies and from the highly interesting material. However, they noted that the courses promoted little social contact with students from other schools. In line with the generally positive feedback on the courses, most of the students said they will recommend the courses to their friends and would like to continue learning in distance learning courses next year.

## 5. Where Do We Go From Here?

As the results of the pilot demonstrate, with adequate individualized support, distance learning can provide a challenging, yet rewarding independent learning experience to gifted students, especially to those in the periphery who do not have access to many other resources. In line with the objectives of the courses, further consideration is needed on how to enhance intellectual meeting and team work among students from different schools who communicate via the internet. In line with the findings concerning in-depth study, another issue to consider is whether to promote more exposure to various content areas, or to in-depth investigation in specific areas of interest, supported by mentors. The focus on these issues will enable the Department of Gifted Students to continue to modify distance learning in an effort to provide an exciting opportunity for gifted and talented students to fulfill their potential in the sciences utilizing different learning styles and world class resources and media.

**References:**


[1] A. Salant, Distance Learning Via the Internet. http://www.amalnet.k12.il/madatec/print/B4_00001.htm. 2003. (in Hebrew).

[2] P. Dallas and M. Franklin, Teaching to the Camera: Learning Long Distance. Paper Presented at the Annual Meeting of the Conference on College, Composition and Communication. (ERIC ED 402604). Milwaukee. 1996.

[3] M. Fleming, T. McCormick, N. Tushnet, and C. Naida. EquityIissues in the Star Schools. Paper Presented at the Annual Meeting of the American Educational Research Association. (ERIC ED 385225), San Francisco, April 1995.

[4] C. Dede, Emerging Technologies in Distance Education for Business, Journal of Education for Business, 71(4) (March-April 1996) 197-204.

[5] A.H. Duin and R. Archee, Collaboraion Via E-mail and Internet Relay chat: Understanding Time and Technology, Technical Communication 41 (1996) 695-708.

[6] G. Kearsley and W. Lynch, Structural Issues in Distance Education, Journal of Education for business 71 (1996) 191-195.

[7] J. Repman and S. Logan, Interactions At a Distance: Possible Barriers and Collaborative Solutions, Techtrends, (November- December, 1996) 35-38.

[8] P. Vicky, Online Universities Teach Knowledge Beyond the Books, HR magazine 43 (July, 1998) 120-124.

*Science Education: Best Practices of Research Training for Students under 21*
*P. Csermely et al. (Eds.)*
*IOS Press, 2005*




# Preparing for Tomorrow by Involving High School Scientists in Leading Edge Research on a National Level

**Joan M. MESSER**
*American Junior Academy of Science*
*900 South Court Street, Ellisville, MS 39437 USA*
*joan.messer@jcjc.edu*

**Abstract**. This paper will first explain the history and philosophy of the American Junior Academy of Sciences (Am JAS). We wil give the results of a national teacher survey that discusses how student projects are chosen, why doing a project gives a student an advantage, and finally give a brief comparison between American students and students from other countries.

The American Junior Academy of Science is composed of America's brightest high school science research students. Selection of these "best of the best" researchers is based on statewide selection of scientific research projects whose quality and sophistication are usually remarkable. Students are selected from the forty-four affiliated state senior academies to be honored and represent their state's organization at the national level. Several million students compete at local, regional, and state levels of each participating state academy with about one hundred fifty chosen to represent their state at the national AmJAS convention.

AmJAS (http://www.amjas.org) is a talented and vigorous organization sponsored by the National Association of Academies of Science (NAAS; http://astro.physics.sc.edu/NAAS/), the national parent organization of state academies and an affiliate of the American Association for the Advancement of Science (AAAS; http://www.aaas.org), the publisher of *Science*. AmJAS student delegates share their research at both the AmJAS and AAAS scientific community.

Since AJAS, NAAS and AAAS annual meetings are concurrent, student delegates have an unparalleled opportunity to meet and interact with scientists from many different disciplines and countries. *For example*, AmJAS delegates were on hand in San Francisco in 2001 when Francis Collins formally released the sequence of the human genome (http://www.jcjc.edu/org/ajas/archives.

As part of the AAAS meetings, AmJAS delegates give both oral and poster presentations of their award winning research projects. They attend AAAS scientific sessions, and tour university campuses and historical sites. One important highlight is always the annual "Breakfast with Scientists," a morning roundtable discussion of educational opportunities and career objectives with notable scientists. Breakfast invitations are extended to scientists attending the NAAS and AAAS meetings and faculty at nearby universities. More than ten Nobel laureates have attended the breakfast in the past three years. Student delegates may talk with scientists with similar interests, as well as with those in different fields of study. The climax of the five-day convention is the awards banquet; a formal event at which each student and chaperone is awarded a certificate and state delegation photographs are taken. Delegates leave the convention having formed a lasting global network of friendships with other high ability and very motivated young scientists.

AmJAS has no income; therefore NAAS board members solicit commercial sponsorships while the state academies or the delegates raise their own funds. Connections with potential sponsors have been made through mutual contacts. An example of this happened in 2001. Harvard University was the AmJAS educational host. Our Harvard hosts mentioned a pharmaceutical company, AstraZenca, which they thought might be interested in supporting our events. Information was sent to a specific company vice president. Experience has shown us that it is much better to contact a specific person. After AstraZenca agreed to support the “Breakfast with Scientists,” the vice president himself attended the breakfast where he met our young scientists. He left that breakfast quite convinced of the value of their sponsorship. Needless to say, AstraZenca has been on board as a faithful sponsor every since.

Some sponsorship has been established during the AAAS/AmJAS convention. As student researchers display their posters on the same floor with AAAS commercial exhibits, potential sponsors view the posters and see for themselves the caliber of these young scientists. Most conventions will find the NAAS board talking with the various convention vendors about AmJAS students. It really helps that these potential sponsors have seen for themselves the outstanding research of AmJAS student delegates.

Each year about 120-175 students and 75 teachers/chaperones attend the AmJAS convention. These students are unique in that they have developed research projects that were judged by senior scientists to be of merit. More importantly, often their work was guided under the auspices of their science teacher(s) and/or senior scientist mentors. How did these students decide on a research topic? How do they establish contact with senior scientist? Here is a brief overview of the process in the United States.

The hardest step is the first step, that is, to decide on a research project. There are two main components, the student's and the instructor's roles. To help decide on a project, students may ask themselves questions such as:

- What are my interests? Hobbies?
- How do I like to spend my time? How does my family spend their time? *(The best projects come from things that perplex students, or their family.)*
- What do I want to be?
- Do I enjoy gathering data inside a lab or in the great outdoors?
- What access do I to research equipment?

If you have limited access to technological research equipment, then look at other avenues for a project. Here are a few hints:

- Look for raw research data, and develop ways to interpret or manipulate it.
- Look at your life. What seems strange? What needs improvements?
- What do you have for research resources? High tech equipment is not needed for math, computer, and theoretical or behavioral biology projects. *For example*, study indigenous animals to understand which plants they use repeatedly, but in limited quantities. These plants may provide some medical benefit.
- Do you already have a science fair project? Can it be extended?

Instructors can foster and encourage students by:

- Requiring that students gather research ideas from television, movies, and newspapers. Internet sites, scientific journals, or magazines often catch a student's imagination. The instructor can help to locate scientists to discuss the feasibility of extending their students' imaginative research idea.
- Meeting in "brainstorming" sessions or round table discussions about research ideas. Many times the best projects come from questions that the students and/or their family are involved in or are perplexed by.
- Contacting potential mentors for a list of research questions that would enhance the mentor's research area. Mentoring is a recruitment tool for colleges and universities. Therefore, most American institutions have incentives for science professors to actively engage in research interactions with young people.
- Checking resources available for schools. Kinston High School (South Carolina) participated in an "adopt-a school" though the Peace Corp. Kinston students compared local data (i.e. pH, temperature, salinity pollutants, species diversity) from a local river to a river located on a Caribbean island.
- Contacting local and/or school machine/fabrication shops to build needed research equipment.

Students and instructors must spend much time one-on-one to develop factual knowledge, lab techniques, *etc.* Many states have courses in which research is part of the core requirement. Some states have "Math and Science" high schools. In other states, instructors spend extra time beyond the normal class time, most often without financial compensation. Some instructors and students have to pay their own expenses if they are selected to move to state or national competitions (AmJAS conventions), but the school's sports team involved in state and national games would not.

So what are the rewards? Deep within the heart of many of American teachers is the love of breaking traditional rote learning. The rich educational experience of science research entices the young researcher into a new and exciting world of science. Traditional science relies on memorization of an incredible amount of trivial information; research makes that information come alive in the students. By far, more students with research experience are accepted into science programs at major universities than uninvolved students. I'd like to tell the story of a young average "C" student from Kentucky. This young man found a research topic that he really liked. He was chosen to represent Kentucky at numerous state and national conferences. He is now an "A" student that will attend a major US university to study science. Another young scientist developed a measuring instrument that is available for purchase from "Frey Scientific," a well known scientific supply house.

Once the students are chosen by their state academies to attend AmJAS, they are eligible to apply to attended international science camp. Four US high school students and I attend the International Science and Engineering Camp 2004 (ISEC 2004) in Pohang, South Korea. By far, the US students had more research lab experience. Preliminary reports from APEC (Beijing, China) this past summer indicate a similar situation among the counties represented at that camp. Individuals on the US teams typically spent two to three years on their individual research topics. The major contributing factor to the American students' research experience is either attendance at "Math and Science" high schools, or their dedicated science teachers electing to make inquiry-based research a graded component of their courses.

The situation in the US is in contrast to the reports given by groups of international educators at ISEC 2004. Asian educators reported that many of their students study in "accessory schools" until late in the evenings. Students are severely limited in their access to research labs. College professors have little incentive to work with secondary school students. Teachers were not trained to assist student with research or to how to encourage the creativity that is necessary in pursuit of an active research project. Students, also, seemed reluctant to pursue a research project by themselves.

Clearly, mentoring high school researchers is a time-consuming process. There are not as many American teachers teaching research-based science courses as in the past. The American undergraduate college curriculum for high school educators does not include "how –to" techniques on mentoring students. This is a factor in declining number of projects submitted to numerous competitive regional, state and national science fairs. Each year the nucleus of research involved students across the U.S. gets smaller and smaller. Even on the college level, there are great concerns about students entering the sciences. The effect transcends to graduate schools where positions are being filled with students from abroad.

AmJAS is looking to the future realizing that scientific advances are becoming more global in nature. We are establishing international ties with Japan, Israel (Weizmann Institute), China, and Korea. We invite collaboration with the nations and institutions that are represented at this conference. We realize that tomorrow's realities are the dreams that begin today. Our dream is that, world-wide, all educators will be better equipped to lead future young scientists into the profession of science. AmJAS wants to have a role in encouraging those young scientists in making a career in science a reality in their lives.

**Acknowledgements**

Special thanks goes to Jerry Guo (SC), Dorcas Green (SC), Nevin Longenecker (IN), Marcia Gillette (IN), Glenn Zwanzig (KY ), Rose Hemphill (OR), Patricia Zalo (FL), Mel Stephen (MA), Peter Langley (OR), and Sister Mary Ethel Parrott.

P. Csermely et al. (Eds.)
IOS Press, 2005


# Shrinking the World by Expanding Research Opportunity

**Peggy CONNOLLY**
*Illinois Mathematics and Science Academy*
*1500 W. Sullivan Rd. Aurora, IL 60506 USA*
*connolly@imsa.edu*

**Abstract**. The Network of Youth Excellence (NYEX) was established to develop scientific talent by promoting and improving international research collaboration among young scientists. With the help of its research partners, the Illinois Mathematics and Science Academy (IMSA) offered opportunities to student members of the Network to work with scientists in the United States. This was a small initiative, but a big success that leveraged existing resources to develop opportunities for international research collaborations.

## Introduction

Venus sprang whole, in all her splendor, from the head of Zeus. In contrast, the great accomplishments of mere mortals do not burst upon the world spontaneously and whole. Whether discovering a new star, identifying a novel immunology technique, or establishing a new international organization, accomplishment proceeds step by step. The goals of the Network of Youth Excellence (NYEX) are significant and noble: fulfilling them will be an enormous task. NYEX will succeed, and succeed it will, by doing many small things well. Small does not mean inconsequential. The challenges for which science seeks solutions are global: eradicating disease, eliminating starvation, rebuilding the environment, developing renewal energy, exploring unknown micro- and macro-universes, ensuring personal safety and international security. Nurturing the next generation of scientists is critical in addressing these challenges, and developing a culture of worldwide scientific collaboration is essential to expeditious progress and the survival of humankind. Success will be measured by small, significant, continual gains. NYEX will evolve and grow stronger as we learn from each other how to develop innovative collaborations.

As a member of NYEX, the Illinois Mathematics and Science Academy (IMSA) was eager to contribute to the organization's goals. Following the 2002 NATO Advanced Research Workshop, IMSA hosted three NYEX students who worked with researchmentors at Fermi National Accelerator Laboratory, Loyola University Medical Center, and Northwestern University. This one small step offers a model for future collaborations

## 1. The Illinois Mathematics and Science Academy Mentorship Program

IMSA is a three-year public residential high school for students talented in science and mathematics. Located near Chicago, it offers one of the most rigorous high school academic programs in the United States. Student who are accepted by IMSA have demonstrated exceptional academic abilities and potential. In addition to college prep and university level classes, IMSA offers a voluntary Mentorship Program in which students participate in research side-by-side with some of the area's leading researchers. Over one hundred institution and several hundred individual mentors have partnered with IMSA in the mentorship program, believing that if we are to have the scientists and scholars we need in the future, it is necessary to encourage students while their interest is keen, to provide them with the opportunity to develop and share knowledge gained through research for the benefit of others and the advancement of society.

Although IMSA is situated in a near-ideal location for research, with Chicago universities, medical centers, museums, and corporations generously welcoming our students, it is not perfect. Students have up to a three-hour commute each way between IMSA and their mentor site. On average of three Wednesdays a month, students have no class obligations and are free to work with their mentors, limiting research to about 20 days during the school year. Despite this schedule and due to their hard work and their mentors' commitment, students have achieved significant outcomes such as publishing in peer-reviewed journals, presenting papers at national and international conferences, and winning research competitions. These successes reflect the many students who work on their projects during vacations, on weekends, and during IMSA's Intersession. It was during Intersession 2003 that NYEX students came to the United States to work with IMSA mentors.

## 2. Intersession 2003

IMSA's winter term begins with a weeklong Intersession. Intersession is an island of opportunity for intensive pursuit of a personal interest, without the distractions and obligations of other classes. Students choose from a menu of activities such as classes outside the regular curriculum, educational travel, and research. This concentrated period of time is perfect for scientific and scholarly projects, and for inviting students from other countries to experience the research environment in the United States. In January 2003, Elza, Kriszta, and Zsolt from the Hungarian Research Student Association spent two weeks in the Chicago area. The first week was devoted to experiencing Chicago: the second, to research.

Four key factors were critical in creating this opportunity, and ensuring its success: the use of existing resources, controlled costs, community support, and student selection.

IMSA thrives within a network of supportive resources. Most importantly in this initiative, IMSA mentors welcomed the students into their laboratories: Dr. Charles Webber of Loyola University Medical Center mentored Elza in immunology research, Zsolt worked with Dr. Don Edward at Fermi National Accelerator Laboratory in physics, and Dr. Zoltan Oltvai hosted Kriszta at Northwestern University Medical School in the Department of Pathology. . Prior to the visit, students and mentors communicated through the Internet, and mentors provided students with reading materials related to their projects. The mentors and their facilities provided materials and research supplies.

Costs were limited to the price of airline tickets. While the cost of tickets is not insubstantial, the absence of other costs for lodging, meals, and transportation made this experience reasonably affordable. IMSA students who were off-campus during the week of Intersession offered their rooms, so IMSA dormitories provided free housing for the three students. The food service provided meals at no cost. Buses chartered by the mentorship program provided transportation between IMSA and the mentors' facilities.

The external community also contributed to the success of the visit. The week prior to Intersession, the students stayed with a family in the community and were able to visit the highlights of Chicago. The Field Museum of Natural History, the Art institute of Chicago, and the Adler Planetarium provided complementary entrance to the students and their hosts. Before starting their research, students and mentors met each other informally over brunch. IMSA students organized some activities during the evenings. Many people contributed to this experience and ensured its success.

A fourth factor in the success of the initiative was the student selection. Each of the students was appreciative of what the mentors provided, gracious and responsible at all times, and flexible and understanding when things didn't go perfectly. IMSA required one of the delegation to be at least 21 years of age and willing to assume responsibility for the students. Elza, a medical student, filled this role, and also had the opportunity to master a novel immunology technique that she brought back to Hungary to teach others. Having these students with us for two weeks was important in developing a working relationship within the international scientific community. On a personal level, it was absolutely delightful hosting Elza, Kriszta, and Zsolt!

The following summer, a reciprocal visit was made by an IMSA student to participate in Hungary's summer science camp for top high school researchers. The hosts covered all expenses in Hungary, so again cost was limited to airfare. Because of her visit, Jeanie strengthened working relationships between IMSA and NYEX, resulting in a high level of interest among IMSA students to participate in NYEX activities. IMSA student are working on contributions to the NYEX Website including interviews with scientists, sample protocols, and research products. Students meet biweekly at IMSA to discuss progress on NYEX and WAYS (World Academy of Young Scientists), the NYEX partner organization for young scientists beyond high school.

## 3. Next Steps

Where do we go from here? Science wears a cloak of mystery, creating the perception that is it too complicated for ordinary people to participate in or even to understand. Nothing could be further from the truth: science is all around us, and we use it every day to make sense of our world and our place in it. Traditionally, the opportunity to do research was reserved for graduate students, or infrequently for exceptional university students nearing the end of their undergraduate careers. Opening the door to young people and inviting them to participate as colleagues in research makes evident the reality that science belongs to everyone and everyone has a place in science. Traditionally, research has been competitive, closely guarded to protect profits, careers, strategic advantage, proprietary information, and national interests. There is an urgent need for scientists to collaborate on a global level to find answers to problems that harm all humankind. Self-interest is ultimately self-destructive. People may be able to identify the borders between countries: disease, starvation, toxic pollution, and poverty do not. The chance circumstances that give individuals or nations access to enormous resources do not convey the exclusive right to their use: the human genome is the legacy of all humankind, not just the wealthy; access to scientific medical care is a fundamental human right, not a privilege for the few; the knowledge of how to exploit the environment does not convey the right to extract its treasures for private benefit or to destroy it for short-term gain.

When eager young scientists and scholars begin serious research in high school, what discoveries will be made in those extra six, seven, or eight years they are expanding existing knowledge? When scientists (and policy makers) collaborate rather than compete to solve the world's problem, how much more quickly and safely will solutions be found? When poor countries and their disadvantaged youth become actively involved in the scientific inquiries that address problems in their own communities, how much more will the standard of living improve for all? By nurturing scientific talent, promoting international cooperation, improving existing programs, and introducing scientific initiatives where they do not currently exist, NYEX has the power to change the culture and practice of science.

What small step can you take that will carry NYEX forward with you? Can you identify a scientist to mentor on-line? Can you host a student? Can you raise money? Can you help plan a conference? Can you advise new mentors how to be successful? Can you counsel students so they will be successful? Can you share a protocol on-line to help students without mentors set up a project? Can you help students edit papers for conferences? Can you inspire colleagues and students to become involved?

The Network of Youth Excellence is an organization of unlimited potential. The challenge of the Network of Youth Excellence is that if we leave here and do nothing, nothing will change. The thrill of the Network of Youth Excellence is that it will become what we make of it.

*Science Education: Best Practices of Research Training for Students under 21*
*P. Csermely et al. (Eds.)*
*IOS Press, 2005*


# Recruitment to Howard Hughes Medical Institute/ National Institutes of Health/ Montgomery County Public Schools, Student and Teacher Internship Program

**Sandra R. SHMOOKLER**
*Montgomery County Public Schools*
*Sandra_Shmookler@fc.mcps.k12.md.us*

**Abstract.** The Student and Teacher Internship Program, called STP, is a partnership between the HHMI, the National Institutes of Health (NIH), and MCPS. High school students apply to, spend a full year doing an internship at NIH. Middle and high school teachers are encouraged by their principals and resource teachers to participate in this program to enhance their teaching content knowledge and professional development. Recruitment to this program is done in many stages and is ongoing. Each school year, presentations are made to science teachers, guidance counselors, resource teachers, and high school principals informing them of the program. Previous student presentations are aired on the local MCPS television station to inform the community. Current student and teacher interns are the best source of recruitment for this successful, highly competitive program. Each generation passes on their excitement and sense of privilege to their peers.

The Montgomery County Public Schools (MCPS) is one of the premier United States (U.S.) school systems. It is a Maryland suburb of Washington, D.C. In the 2003–2004 school year, there were 192 schools. Three new schools opened this school year. There are approximately 140,492 students in grades kindergarten through high school.

MCPS is like the United Nations. Children come from all over the world and speak more than 123 native languages. Twenty-five years ago the school system was predominately White. Today there is no majority population. Montgomery County, Maryland covers 550 square miles and is the 18th largest school system in the US.

Many families move to Montgomery County because of the excellent free public education. Needless to say, science is a most important component of the rigorous academic curriculum. MCPS, thanks to generous grants from the Howard Hughes Medical Institute (HHMI), has been able to supplement the science curriculum in all grade levels, including the Student Inquiry Project for elementary students.

A premier high school program is the Student and Teacher Internship Program, called STP. This is a partnership between the HHMI, the National Institutes of Health, and MCPS. NIH is the steward of medical and behavioral research for the Nation, supporting biomedical research on their Bethesda, Maryland, campus and throughout the United States.

Approximately 22 high school students, selected from applicants from 24 high schools, spend a full year doing an internship at NIH. The program begins in the summer with a two-week course in laboratory techniques, taught by MCPS faculty and scientists from the NIH. The students spend the remaining summer in a laboratory at NIH. Selected teachers, who do an eight-week summer internship, are in the same class as students prior to entering their laboratory. Teachers culminate their internship with a presentation about their research and how to apply their experiences to the classroom. Students return to their NIH laboratories in September for half-day—typically spending mornings at school, usually taking four classes and then remainder of the day at NIH. The yearlong experience culminates at a dinner symposium where each student presents his or her own research. Typically, the audience is made up of NIH scientists, student preceptors and mentors, as well as elected officials, parents, teachers, and guests, are always amazed at what these young people have accomplished. Often student's work is so outstanding it is published, and students are invited to present their research at conferences. NIH often features student work in their monthly publications as we ll. Also, there are frequent newspaper articles and schools newsletters that feature student's work and highlight their accomplishments.

Recruitment to this program is done in many stages and is ongoing. In other words, it is an internal component of the program. Because the school system is so large, it is a continuous goal of the program administrators to ensure that qualified students in all high schools have an opportunity to apply. Therefore, at the beginning of each school year, presentations are made to science teachers, guidance counselors, resource teachers, and high school principals informing them of the program. Presentations also are continuously made at parent-teacher association meetings and other local events. Each high school principal can recommend up to four students from their high school to apply. All applications are reviewed once a year by a team of educators and NIH scientists.

As previously stated, the recruitment also includes presentations in each high school by current interns to other students describing their internship activities at the NIH. Each semester current student interns do a short presentation in their school, either in a science class or in the career center. Their presentation is treated like any other outside speaker and advertised in the school. Good recruitment includes assurances that those adults who work with students on a daily basis have a clear understanding of what this internship entails. Thus, they can encourage students to consider applying and also provide further information.

Previous student presentations are aired on the local MCPS television station to inform the community. Each high school resource teacher has a copy to share with students and others.

The STP web site, http://www.mcps.k12.md.us/departments/intern/stp/ is probably the best marketing tool. It provides further detailed information. We receive continuous questions from MCPS parents and students, as well as from around the county. We also are able to help other school systems with internship programs.

Whetting young people's appetite to the joys of science begins early in a young person's life. The elementary program, *Student Inquiry Project*, begins to make science fun. The goal of the program is to provide teachers and students with hands-on inquiry based science curriculum lessons. Teachers encourage children to enjoy their experiments and model the high school program by providing opportunities for students to present their work to other, including classmates and parents.

A middle school program called *Fun with DNA* allows $8^{th}$ and $9^{th}$ grade students to experience a dynamic hands-on three-week summer opportunity, mentored by teachers and high school students. One activity includes DNA work, using their own saliva. This program acquaints students with the various aspects of the field of biotechnology in an effort to nurture their interest in science and is a very popular component of recruitment.

Also, on a middle and high school level, the *Student Academy of Science (SAS)* has been developed in many secondary schools to encourage young people to explore the joys of science. HHMI funds this after-school mentoring program also. Again, this program is to stimulate student interest in career opportunities in biomedical science through presentation by scientists, discussions of videos, and field trips to such places as the National Library of Medicine, and Armed Forces Institute of Pathology.

Thus, though early exposure to science activities and encouraging students to take more science and math helps to prepare them to apply for internships.

*P. Csermely et al. (Eds.)*
*IOS Press, 2005*


# Out of School Science Programs for Talented Students: A Comparison

**Rena F. SUBOTNIK**
*Center for Psychology in Schools and Education*
*American Psychological Association, USA*
*rsubotnik@apa.org*

**Abstract.** Six categories of out of school science programs for talented students are described in relation to six criteria: method of selection, goals, structure, benefits, drawbacks, and predictions about participants' future in science. The context for development of this analysis is described as well as suggested directions for the future.

## 1. Background

The Center for Psychology in Schools and Education (CPSE) at the American Psychological Association promotes the high quality application of psychological science to education programs and policies. The mission of CPSE is to generate public awareness, advocacy, clinical applications, and cutting-edge research to enhance the educational opportunities for students at all levels of schooling. The Center is made up of four offices that enhance the mission of CPSE:

- *Post Doctoral Education*: PhD students in psychology are very well trained in research methodology and in theories of teaching and learning. Their work, however, tends to take place in laboratories or in one-on-one clinical settings. This program matches outstanding new psychology PhDs with experienced education researchers for a two-year apprenticeship conducting research in schools.
- *Optimizing Student Success with Reasoning, Resilience and Responsibility:* These attributes enhance student engagement in school and in academic achievement. A team of researchers and expert public school teachers organized by CPSE

developed a teacher-training program that infuses Reasoning, Resilience, and Responsibility into 3rd grade mandated curriculum in local public schools.
- *Coalition of Psychologists:* Within the discipline of psychology, one can find over 50 sub-disciplines, with often conflicting goals. CPSE hosts a coalition of psychologists from 11 sub-disciplines to focus on common concerns related to teaching and learning in schools.

## 2. Center for Gifted Education Policy

The fourth office in CPSE is the Center for Gifted Education Policy. The following activities define its mission:
- *Continuing Education for Psychologists*: These sessions offer awareness training to psychologists who wish to expand their services to gifted children, adolescent and adult clients.
- *Prototype Talent Development Programs*: The Gifted Center has created two evidence-based out-of-school programs for talented adolescents. The Young Scholars Social Science Summit is a one-day program focused on social science research methods. A second program, the Pinnacle Project, will be described in more detail later in this chapter.
- *Advice to Parents, Teachers, and Psychologists*: This advice is delivered through two channels. One is the CGEP listserv, which includes over 350 participants from around the world. The listserv offers a place to post announcements, queries, and requests for research subjects or good scholarly references (to join contact Jason Gorgia at jgorgia@apa.org). The other channel is through the CGEP website (www.apa.org/ed/cgep.html). In addition to describing the Center's activities, the website lists references to various organizations in our field and the tables of contents for major journals in gifted education.

## 3. Experiences and Background with Out of School Science Programs

Often I am asked for advice from parents, teachers, and psychologists about placement of gifted adolescents in out of school talent development programs. I have developed a set of criteria by which I offer my advice, taking into account the needs of the student and the benefits of the program. The judgments are also based on the following activities I have been intimately connected with.
- *Westinghouse Science Talent Search Study* [1]: A 13-year longitudinal study of students who won the most prestigious science award in the US. The winners were selected on the basis of a research paper and a set of interviews with eminent scientists.
- *Evaluation of Olympiad Studies* [2]: We reviewed studies conducted by scholars around the world regarding the background variables of participants in the Olympiads as well as some measures of the reported effects of Olympiad participation.
- *Pinnacle Project* [3]: Conducted for three years (2001-2004) as an intergenerational, interdisciplinary mentorship program, the Pinnacle project

addressed high level talent in the sciences, and also in the arts and professions such as journalism.

- *Comparison Study* [4]: In collaboration with the College of William and Mary, CGEP explored differences between participants in an out of school research science program and an Olympiad-type program with regard to school and home support and preparation.

## 4. Six Models of Out of School Science Talent Development

From a review of programs offered at the 2004 NATO-UNESCO Advanced Research Workshop on Science Education: Talent Recruitment and Public Understanding, as well as the programs widely promoted to students in the United States, I have identified six categories of out of school science experiences.

- *Kitchen science – conducted informally at home alone or with friends. Usually involves chemistry or rocketry.*
- *Olympiads and other test-based competitions*
- *Science research programs*
- *Intergenerational, interdisciplinary programs*
- *Intensive summer courses*
- *Science clubs*

## 5. Six Dimensions of Analysis

When considering each program for an individual student, I consider the following criteria:

- *What is the selection process for the program?*
- *What are the goals of the program?*
- *What is the structure of the program?*
- *What the main benefits of the program?*
- *What are the main drawbacks of the program?*
- *What predictions might we make about the effects of the program on participants?*

## 6. Brief Analyses by Program:

Kitchen Science:

- *Selection:* Self-selection by an individual on his or her own (with or without family or friends)
- *Goals:* The purpose of Kitchen Science is to fulfill curiosity and to have fun exploring various materials and interactions.
- *Structure:* None
- *Benefits:* The explorations are conducted in the context of every day life and can feed a lifetime of curiosity about scientific matters. Occasionally, a kitchen experiment may lead to an invention that brings profit to a young scientist.
- *Drawbacks:* Unless the activities can draw the attention of a mentor, the depth of scientific exploration can be limited. Further, without a mentor, it is difficult to

find out about career paths, out of school programs and higher education opportunities.

- *Predictions:* Some of the most accomplished scientists began their lives in science in this way. Other kitchen scientists enjoy playing around but lose interest when the more tedious aspects of conducting science, including required academic coursework, take a larger role.

Olympiads and Other Competitions:

- *Selection:* Is conducted on the basis of a series of increasingly more difficult tests
- *Goals:* Capitalizes on the joys and challenges of competition experienced by athletes
- *Structure:* Top scorers on first test are invited to sit for next test. One or two additional tests winnow down participants to approximately 20 who attend a ten day or two week intensive preparation for international competition. Five or six team members are selected for international competition based on performance at the camp.
- *Benefits:* Competition serves as a motivator to learn a tremendous amount of information. Students meet others who enjoy both fast paced competition and the sciences.
- *Drawback:* The practice of preparing for speedy solution of given problems does not provide realistic preparation for scientific careers.
- *Predictions:* Because of the drawback listed above, it is best to also experience a Science Research program in order to get a more realistic sense of science as a career.

Science Research Programs:

- *Selection:* Selection tends to be based on a variety of sources including essays, standardized test scores, and teacher recommendations.
- *Goals:* To serve as an apprenticeship in the techniques, knowledge, and values of science research.
- *Structure:* Intensive involvement in an active professional scientific endeavor. Often relevant coursework is provided as well.
- *Benefits:* Participants learn that scientific efforts are complex and do not always lead to success. They also get a realistic picture of careers in science research as well as socialization into the value system that supports retention in the field.
- *Drawbacks:* Programs select willing mentors who open their labs to adolescent scientists and provide them with attention and support. But too often, the transition into higher education leads to disappointment when lab heads are not interested in collaborating with anyone below graduate school level.
- *Predictions:* If students maintain their interest through the first two years of undergraduate education, they are likely to make a successful transition into science research careers.

Intergenerational/Interdisciplinary Pinnacle Project Model:

- *Selection*: Eminent mentors in seven or eight disciplines in the arts and sciences provide criteria for what they think a talented adolescent should have accomplished at this point.
- *Goals*: Providing an environment and relationships to stimulate intellectual and creative growth leading to path breaking work.
- *Structure*: A summit including meetings with mentors in one's field as well as in other disciplines. Over course of summit and following year, teams conduct intensive work on a project as well as expose one another to path breaking work in other disciplines.
- *Benefits*: An expanded view of giftedness across various disciplines in the arts and sciences as well as a more sophisticated world view borne from exposure to frontiers of various fields and disciplines.
- *Drawbacks*: Funding challenges to support such an elite program. Adults get so excited by the interdisciplinary activities, that adolescent participants can struggle to be heard.
- *Predictions*: Offers a path to eminence with some detours in other fields.

Intensive Summer Courses (contributed by Harald Wagner, Bildung und Begabung e.V., Bonn, Germany):

- *Selection*: Selection is based on a combination of standardized tests and/or school recommendations.
- *Goal:* Provides challenging coursework and social opportunities with like-minded peers and inspiring teachers; fostering autonomous learning, knowledge production and teamwork; improvement of oral presentation techniques and scientific writing.
- *Structure:* Courses of one to several weeks duration provided in diverse academic areas at each site. Extensive leisure time is available for sports, music, excursions, discussions, socializing etc.
- *Benefits:* Long lasting contacts and friendships in an alumni network; realistic assessment of and increased confidence in one's own abilities; improved motivation, cooperativeness and communication skills.
- *Drawbacks:* Relatively expensive to participate, unique "hot house" experience that has no chance for continuation in "real life," sometimes making return to school a painful contrast.
- *Predictions*: Increased academic and social self-confidence provides a valuable bridge from school to university.

Scientific Clubs (Contributed by Tamas Korcsmaros, Hungarian Research Student Association):

- *Selection*: Individuals create a group based on mutual interests in science.
- *Goal:* The clubs allow for sharing scientific knowledge and help in solving problems encountered in members' projects.
- *Structure*: Like any friendships established around mutual interest, the structure can vary from none to a lot, depending on the projects and the personalities of the participants.

- *Benefits*: Provides members with partners for discussing new and creative ideas, dealing with successes as well as disappointing project results, and advice on publication and other desired outcomes.
- *Drawbacks*: The intensity of the relationships can enhance the likelihood of budding romance. When these romances end, the group camaraderie can be affected.
- *Predictions*: Can lead to lifetime friendships that help to maintain vocational or avocational interests in science.

## 8. Future Directions

- The status quo: Maintaining the profusion of models stimulates creative approaches, many options for different students, and exciting findings for use in research on talent development and recruitment.
- It might be useful to develop a compendium of existing programs in each category (and add categories as they emerge) for use by teachers, students and parents.
- We could also take a harder look at the strengths and weaknesses of each program design and make adjustments. Some of the programs have been in place and successful for such a long time that change is not viewed as a valuable pursuit.
- The field of talent development would be much enhanced if we could commission a longitudinal comparison study of the career choices and success of participants of each type of program.
- Generate some proposals for policy makers for out-of-school and in-school talent development and recruitment based on best practice conducted in exemplary programs. Policy makers might consider integrating some of the successful practices from each type of program into the school curriculum and schedule.
- Convene program leaders, funders, and policy makers to explore ways of expanding and sustaining successful models in and out of school.

**References:**

[1] Subotnik, R.F., Maurer, K., & Steiner, C.L. (2001). Tracking the next generation of the scientific elite. *Journal for Secondary Gifted Education. 13*, 33-43. Subotnik, R.F. & Arnold, K.D. (1995). Passing through the gates: Career establishment of talented women scientists. *Roeper Review, 18* (1), 55-61. Subotnik, R.F., Duschl, R., & Selmon, E. (1993) Retention and attrition of science talent: A longitudinal study of Westinghouse Science Talent Search winners. *International Journal of Science Education, 15* (1), 61-72.

[2] Subotnik, R.F., Miserandino, A., & Olszewski-Kubilius, P. (1996). Implications of the Mathematics Olympiad studies for the development of mathematical talent in schools. *International Journal of Educational Research, 25*, 563-573.

[3] Subotnik, R.F. (May, 2003). Through another's eyes: The Pinnacle Project. *Gifted Child Today, 26*, 14-17. Dittmann, M. (2003) The top of their class. Monitor on Psychology, 65-67.

[4] Subotnik, R.F. & VanTassel-Baska, J. (In preparation). Background variables in two contrasting models of highly selective out of schools science programs: A preliminary analysis.

Science Education: Best Practices of Research Training for Students under 21
P. Csermely et al. (Eds.)
IOS Press, 2005



# Science Mentorship Program for High School Students in Korea

**Myoung Hwan KIM**
*Kim Institute for the Korean Gifted and University of Incheon (Dept. Of Physics)*
*1640-29 Seocho-dong, Seocho-gu, Seoul, Republic of Korea,*
*mhkim@thinkid.com*

**Abstract**. Science mentorship programs in Korea were initiated by the Science Gifted Education Center of university from 2000 and the Busan Science Academy from 2002 and Science highschool from 2003 which was sponsored by the Ministry of Science and Technology and the Korea Science and Engineering Foundation(KOSEF). Science mentorship programs of university was started from 1998 and science mentorship programs of the Busan Science Academy was established in 2002. KOSEF gave support to 72 science mentorship programs for science highschool students in 2003 and 79 science mentorship programs in 2004.

## Introduction

The Korean Government, the Ministry of Science and Technolgy (MOST) and the Korea Science and Engineering Foundation (KOSEF), has given support to the Science Gifted Education Centers of universities from 1998 for the purpose of educatiing the gifted elementary and middle school students and now 23 Centers are supported by the Government.

In order to support continuously the science gifted middle school students in the Science Gifted Education Center of university, MOST announced to support one Science highschool among 16 Science highschools and select one Science highschool as Busan Science Academy(BSA) in 2001. In support of MOST and KOSEF, all the students of BSA have to complete personal research of science as the curriculum and they are doing now.

To encourage science gifted education of Science highschools, MOST supported Science highschools from 2003 such as science mentorship programs, teachers' research programs, students' scientific activities and improvement program for science education environment. 292 students had done 72 science mentorship programs in 2003 and 323 students are doing 79 science mentorship programs in 2004.

In this paper, the science mentorship programs for the highschool students in BSA and Science highschools in Korean will be introduced.

## 1. Science Mentorship Program for BSA Students

BSA, officially approved as the school for scientifically gifted students in 2002 by the Ministry of Education and Human Resources (MOE), has the educational objectives to enhance the creativity and scientific research abilities of the students and to promote selfdirected learning abilities that lead to the promotion of new knowledge and to teach the skills and ethical attitudes toward science, required of scientists of world stature.

144 students who are in the 7th, 8th and 9th grade, 8 classes (18 students a class), can enter BSA every year and they can graduate after completing the given course. The curriculum is composed of subjects (145 credits), Research and Education (R&E, science mentorship program, 30 credits) and extracurricula activities. Research and Education is the other name of science mentorship program and which is composed of personal research and commit education.

**Table 1. Curriculum of Busan Science Academy.**

| Subjects Areas | | | Curriculum (The minimum grade credit) | | | |
|---|---|---|---|---|---|---|
| | | | Common Basic | Elective Courses | | Total |
| | | | | Basic Elective | Intensive Elective | |
| Subject Areas | General Subjects | Korean Language | 8 | 4 | 0 | 12 |
| | | Social Studies | 6 | 12 | | 18 |
| | | Foreign Language | 8 | 6 + 4 | | 18 |
| | | Art and Physical Education | 10 | 2 | | 12 |
| | | Sub-total | 32 | 28 | 0 | 60 |
| | Major Subjects | Mathematics | 10 | 6 | 36 | 85 |
| | | Science | 20 | 4 | | |
| | | Information Science | 4 | 2 | | |
| | | High Technology | 0 | 3 | | |
| | | Sub-total | 34 | 15 | 36 | 85 |
| Total Subjects Grade Credit | | | 66 | 43 | 36 | 145 |
| Personal Research | | | 20 | | | 20 |
| Commit Education | | | 10 | | | 10 |
| Grand Total units | | | 96 | 79 | | 175 |
| Extracurricular Activities | Club Activities | | over 40 hours every year | | | |
| | Social-Service | | over 40 hours every year | | | |

□ 1 credit is composed of 50 minutes and one semester is composed of 16 weeks. 1 credit of Experiment is composed of 100 minutes.

□ Physical education should be done every semester.

The Reseach and Education programs in 2003 and 2004 are indicated in the table 2.

**Table 2. Numbers of Participants and Subjects in science mentorship program for Busan Science Academy Students.**

| | Students | Math | Physics | Chemistry | Biology | Earth S. | Info. Sci. | Total |
|---|---|---|---|---|---|---|---|---|
| **2003** | 144 | 6 | 7 | 5 | 8 | 5 | 4 | 35 |
| **2004** | 288 | 5 | 21 | 8 | 14 | 2 | 5 | 55 |

## 2. Science Mentorship Program for Science High School Students

The first science highschool, Kyunggi Science highschool, had opened in 1983 and 16 science highschools in 2004 selected about 1,200 students (53classes, 23students a class) at the ratio 0.2% among 0.6 million students of the same grade.

Science mentorship programs for science highschool students have been begun as after school program since 2003 in support of MOST and KOSEF.

The theme of science mentorship program is selected in two ways. One is that professors suggest the themes and students select among them, the other is that students determine a theme they want to study and ask a professor to be a mentor. The charactoristics of science mentorship program are followings.

- A period of science mentorship program is in one year (from July to Feb. in 2003, from May to Feb. 2004)
- The students have to submit their activity record every month.
- A mentor, Professor, has to submit his/her instruction record every 3 months.
- A mentor or an academic advisor has to evaluate the students and submit the results every 3 months.
- After carring out the program, the students have to express the results of their research.

The Science Mentorship Programs in 2003 and 2004 are indicated in the table 3.

**Table 3. Numbers of Participants and Subjects in science mentorship programs for Science High School Students.**

| | Students | Math | Physics | Chemistry | Biology | Earth S. | Info. Sci. | Total |
|---|---|---|---|---|---|---|---|---|
| **2003** | 292 | 18 | 13 | 14 | 13 | 9 | 5 | 72 |
| **2004** | 323 | 20 | 14 | 14 | 14 | 10 | 7 | 79 |

16 researches, as a list in table 4, were selected with honors among 72 researches accomplished in 2003.

**Table 4. A list of excellent evaluation in 2003 science mentorship programsfor Science High School Students.**

| Subject | Title |
|---|---|
| Math | A Study on the Geometrical Representation and It's Application of 4-Dimensional Figure |
| Math | A Study of the Factorization with Elliptic Curves |
| Math | Discovering mathematical structures through symmetries of spaces |
| Math | Population dynamics in mathematical modeling |
| Informatics | New String Matching Algorithms for Intrusion Detection and Response System |
| Physics | Development and characterization of non-contacting viscosity device of liquid |
| Physics | Visualization of acoustic interference and diffraction by measurement using microphone array and computer interface |
| Earth Sci. | Analyses of Planetary Satellite Orbits Observed with Digital Camera |
| Earth Sci. | Characteristics of Earth Gravity and Gravity Survey |
| Chemistry | Development of New Blue Light Emitting Material for Next Generation Display through Creative Scientific Investigation |
| Chemistry | Understanding of Spectroscopy and Its Application to the Characteristics of Nanoparticles |
| Chemistry | Studies on the vital enzymes using the recombinant DNA |
| Biology | Analysis of activation structure in$\alpha X\beta 2$ integrin involved in leukocyte migration |
| Biology | Action mechanism of nuclear hormone receptor in Yeast |
| Biology | Cell fusion and nuclear movement of wound-healing. |
| Biology | Understanding genetic concept through production of double mutation between multi sex combs and pleio-homeotic gene, and its effects on fly development |

### 3. Reponses of Science Mentorship Program

The participants in 2003 responded to our questionaires as follows. The period of science mentorship program in a year was 26 days (95 hours). The ratio of education and experiment was 2:3. 61.1% of the students, 60.5% of the teachers and 46.8% of mentors thought the subjects were difficult to the abilities of the students.

The degree of the satisfaction about science mentorship program was as follows. (very satisfaction 5, very dissatisfaction 1):

- Experiencing in the experiment facilities of university : 3.8
- Activities of science research at high level : 4.2
- Teaching of university professor : 4.4
- Helpful to university entrance : 4.2
- Having a chance of enrichment about science lessons at the school : 3.8
- Helpful to school science lessons : 3.3

50.5% of the students had a clear memory of experimental experience and using the high technical equipment, 25.0% of the students had a clear memory of meeting the professors and Ph. D., the others mentioned attitude of research and the course to take after graduation.

Teachers and mentors evaluated that 84.7% and 74.9% of the students carried out their research actively and 81.7% and 78.0% of them were satisfacted to the science mentorship program.

Student 1: **"It is good to me that I experienced the activities of practical research with the professor."**
Student 2: **"I am happy to have a chance that I could discuss the task with professor profoundly."**
Student 3: **"I am satisfied that I could operate and make the use of the equipments at the university or institute such as NMR, SEM, TEM."**
Student 4: **"I am intersted in the new field of science unfamiliar to the high school lesson."**
Student 5: **"I knew how to applicate what I learned in the school.**"

Professor 1: **"I think it is important of the highschool students to take the sense and environment about scientific researches."**
Professor 2: **"I think the students could have confidence in themselves and abilities to complete the project as a member of research group."**
Professor 3: **"It is helpful of the students to have a chance to experience the theme and the problems and the tendency in recent academic circles."**
Professor 4: **"I think the students could enhance their abilities of challenging the new problems and how to solve them."**

### 4. The closing

After introducing the science mentorship program in Korea many scientists and professors in the field of science joined the program and concerned in science education of highschool students, especilly scientifically gifted students. About 600 students and 140 mentors in the field of science participate the science mentorship program in 2004 and the more in next year.

We have a project to publish 'Young Scientists Journal' and annual conference for Young Scientists to give a chance to express the results of science mentorship programs.

**References:**

[1] M. Kim. A Presentation and Evaluation on Final Results of Research and Education Program - Final Report. 2003.
[2] M. Kim. A creative research project using mentorship for science high school. 2004.

# Education of Gifted Elementary and Middle School Students with University Faculty in Korea

**Byunghoon CHUNG**
*Science Education Institute for the Gifted and Talented,*
*Cheongju National University of Education,*
*Cheongju, Chungbuk, 361-712 Korea*
*bhchung@sugok.cje.ac.kr*

**Abstract**. The MOST and KOSEF initiated and have supported science education for the gifted and talented in Korea. Nineteen local institutes for the gifted education had been established at the universities since 1998. The institute identifies the giftedness of elementary and middle school children and provides them with extra curricula in science, mathematics and informatics. Such programs have been accelerated by legislation of the Gifted and Talented Education Promotion Law in 2001. The KOSEF had financed ca. 8.2 million Euro and 13,021 students had completed the gifted education program.

## Introduction

The MOST (Ministry of Science and Technology) and MOE (Ministry of Education and Human Resources Development) support gifted and talented science education in Korea, even though their policies and conceptions are different from each other in many aspects. The MOST aims to bring a very able student up to be the highly qualified scientist and expects Korea to be one of the top ranked countries with the high tech industry, while the MOE considers that the gifted education is a part of encouraging individual development of students within the whole education system.

In the year of 1998 the MOST launched out the science education projects for the gifted and talented through incepting its own organization in the KOSEF (Korea Science and Engineering Foundation). The KOSEF has accordingly established and financially supported the educational institutes of the university which should be responsible for the local gifted educations. The institutes recruit and identify the very able pupils in science and mathematics, and provide them a certain amount of appropriate learning courses. The projects of MOST for promoting gifted and talented education cover the following categories:

- Supporting the Institutes for the Gifted and Talented(ITG)
- Busan City Science Academy Project
- Science High Schools Project
- Research Institute for the Gifted and Talented in Korea Advanced Institute of Science and Technology (KAIST)
- President Science Scholarship

Since the Gifted and Talented Education Promotion Law (GTEP Law) in 2001 and its enforcement ordinance in the next year had passed the National Assembly, gifted education in Korea came to be completely equipped with a legal, administrative, financial and finally organizational basis.

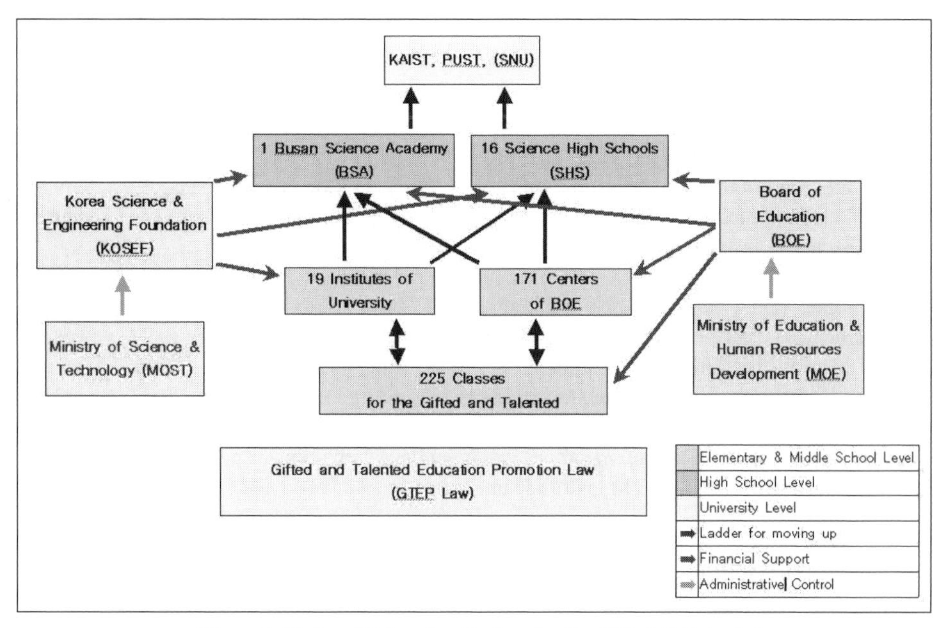


**Figure 1. National Education System for the Gifted & Talented.**

The gifted education has thereafter expanded rapidly in public area, especially in science and mathematics subjects, while the lack of specialists and qualified teachers has became an impediment of the gifted education. In this paper I will give an overview of the status of the IGTs.

## 1. Institute of Science Education for the Gifted and Talented: IGT

The MOST and KOSEF have, since 1998, established total 19 Institutes of Science Education for the Gifted and Talented (IGTs) at the universities [1]. The IGT is responsible for the local educational service in its province and metropolis. They recruit the very able elementary and middle school students and identify their giftedness in subject area including math, science and informatics. The identification process in the IGT usually goes through multiple stages of selection, i.e. the recommendation evaluation, two steps of written tests and lastly the practical skill test on experiment ability. The IGT provides one year course of the extra curricular which is run as weekend program during the term, while the IGT students also enjoy a science camp or an intensive course during school vocation. Some of the IGTs offer additionally a cyber courses in the website. About 100 to 150 lesson hours except cyber course are usually offered per year. During the first year course (Basics Course: BC) the problem solving ability and creativity of students should be observed and evaluated. We perceive the student of BC as a 'potentially gifted student'.

Students should be reselected at the end of BC; about a half of students can continue the second year course, namely the Advanced Course (AC). Students of elementary school level who show the most possible giftedness will move up to the upper level without a screening process. At the end of AC in the middle school level only few students identified as gifted and talented can have a chance to get a one to one mentorship, so they can study at least 7 years long at IGT.

The contents focus on the hands on experiments in BC. Those of AC consist, however, of either the theoretical and abstract form, for example, just like a Gedanken-experiment (though experiment), or of the long term problem solving (i.e. project type activities). Most of lecturers are university professors, but the Ph. D. experts or qualified school teachers are engaged also in teaching. Teaching staffs should develop a teaching program by him/herself and revise it after lesson.

Some IGTs, for instance Cheongju IGT have introduced, in addition to the achievement test, an observational assessment system by assistant teacher (AT) which can help to identify the giftedness of students. Every learning group (usually of 5 to 7 students) has an AT who only observes the students behavior during solving the problem, but neither helps them, nor gives them instruction. After class work the AT hands in the recorded checklist and the written report about each student behavior, especially about his/her creativity, problem solving ability, attitude and so on.

**Table 1. Scheme of Courses.**

<table>
<tr><th>Level</th><th>Classes</th><th>Course</th><th>Term</th></tr>
<tr><td rowspan="2">Elementary School (5-6th Grade)</td><td rowspan="2">Math, Science, Informatics</td><td>Basic</td><td>1 year</td></tr>
<tr><td>Advanced</td><td>1 year</td></tr>
<tr><td rowspan="2">Middle School (7-8th Grade)</td><td rowspan="2">Math, Physics, Chemistry, Biology, Earth Science, Informatics</td><td>Basic</td><td>1 year</td></tr>
<tr><td>Advanced</td><td>1 year</td></tr>
<tr><td>Mentoring (no limit for Grades)</td><td>Not specified</td><td></td><td>Usually 1 year</td></tr>
</table>

## 2. Status and Finance of IGT

The nineteen IGTs are founded by taking the geographical condition and population into consideration. A brief information about the IGTs is shown in the table 2.

**Table 2. IGTs in Korea.**

| Area | | University/Website | Foundation Year | Level offered |
|---|---|---|---|---|
| Metropolis | Seoul | Seoul National University http://gifted.snu.ac.kr | 1998 | M* |
| | | Yonsei University http://tobit.yonsei.ac.kr/~gifted/ | 1999 | M |
| | | Seoul National University of Education http://www.sanion.com/science/index.html | 2000 | E** |
| | Incheon | Incheon University http://isep.incheon.ac.kr | 1998 | E/M*** |
| | Busan | Pusan National University http://gifted.pusan.ac.kr | 1999 | E/M |
| | Daejoen | Chungnam National University http://gifted.cnu.ac.kr | 2003 | E/M |
| | Kwangju | Chonnam National University http://csge.chonnam.ac.kr | 1998 | E/M |
| | Daegu | Kyeongbuk National University http://www.secgy.org | 1998 | E/M |
| | Ulsan | Ulsan University http://gifted.ulsan.ac.kr | 2003 | E/M |
| Province | Kyeonggi | Ajou University http://cge.ajou.ac.kr | 1998 | E/M |
| | Kyungnam | Kyeongnam University http://nobel.kyungnam.ac.kr | 1998 | E/M |
| | Kyeongbuk | Andong National University http://anu.andong.ac.kr/~gifted/ | 2003 | E/M |
| | Kangwon | Kangwon National University http://www.kangwon.ac.kr | 1999 | E/M |
| | | Kangnung National University http://gifted.kangnung.ac.kr/ | 2000 | E/M |
| | Chonnam | Sunchon National University http://secgy.sunchon.ac.kr | 2003 | E/M |
| | Chonbuk | Chonbuk National University http://gifted.chonbuk.ac.kr | 1998 | E/M |
| | Chungnam | Kongju National university http://gifted.kongju.ac.kr | 2000 | E/M |
| | Chungbuk | Cheongju National University of Education http://www.cucocr.org | 1998 | E/M |
| | Cheju | Cheju National University http://gifted.cheju.ac.kr | 2000 | E/M |

* M: Middle School Level
** E: Elementary School Level
*** E/M: Elementary and Middle School Level

The KOSEF organize an IGT evaluation committee which grades yearly the project result of each IGT. IGT has to submit the annual report and the developed teaching materials before the evaluation. The evaluation puts emphasis mainly on the contents of program, the quality and personal sincerity of teaching staffs, and student management. All of these are inevitable for effective education for the gifted and talented. The evaluation result will have influence on amount of financial aid.

The budget of IGT consists of the financial aid from KOSEF and local support. The former is given much weight in the total IGT expenditure, varies with the IGT evaluation result, size of IGT and business years. In 2003 the average financial support per IGT was about 112 thousand Euro [2]. The latter, however, depends on the university policy and regional patronage. It varies IGT by IGT from zero to several hundred thousand Euro. The major parts of expenditure of IGT are personal expenses (lecture fee and expert pay), research expenses and material costs. No tuition fee is therefore charged from parents.

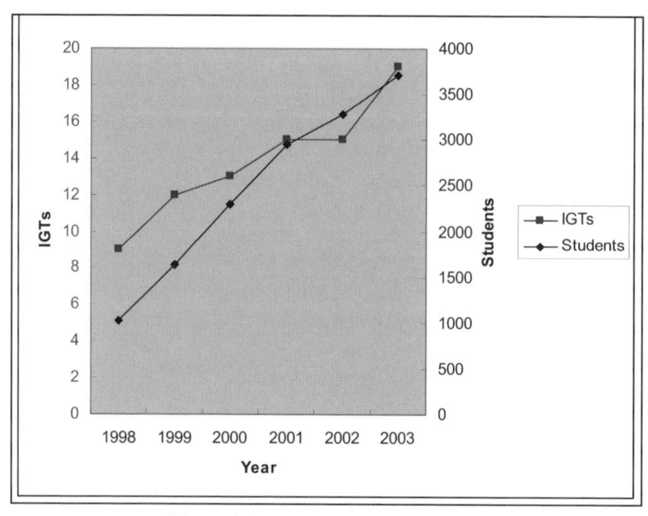


**Figure 2. Number of IGTs and students.**

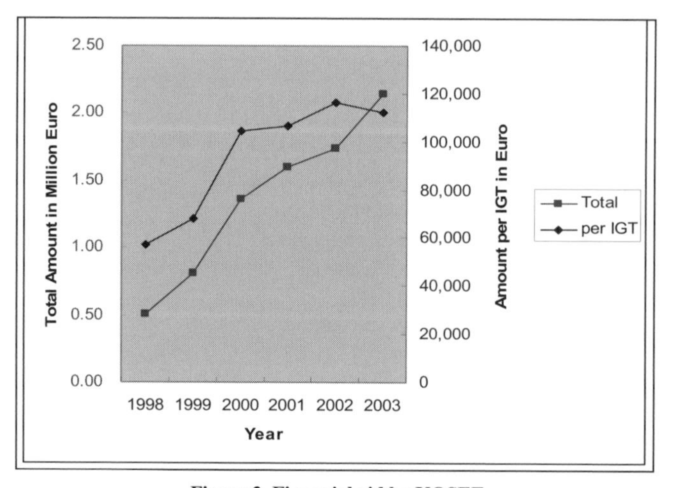


**Figure 3. Financial aid by KOSEF.**

## 3. Conclusion: Perspectives and Expectation

Seven year experiences of IGT may not be enough to realize an ideal system for gifted education. There are also problems to be overcome: professors participating in the project always suffer from lack of time for their own research work. Teaching the gifted would not be meaningful without studying the gifted as well. Overlapping and competition in public system between MOST and MOE will reduce the effective promotion of gifted education.

Nevertheless the initiative of MOST, KOSEF and IGTs have obviously contributed to developing and expanding the public sector for gifted and talented education. The greater part of students of IGT go to either Busan Science Academy (this is the only school for the gifted which has a legal basis) or local Science High Schools. We expect that those who have studied in IGT will become the highly competitive scientists in the world.

**References:**

[1] Korea Educational Development Institute, Current Status of Teaching and Learning at Educational Institutions for the Gifted. KEDI, Seoul, 2003.

[2] Science Gifted Education Institute of Cheongju National University of Education (CNUE), The 2003 Annual Report of the Science Gifted Education Institute of CNUE, CNUE, Chengju, 2004.

# Session IV

# Successful Practices of Research Training – Western Europe

*Science Education: Best Practices of Research Training for Students under 21*
*P. Csermely et al. (Eds.)*
*IOS Press, 2005*



# Life Sciences in Austrian Schools: How Do We Get There?

**Susanne SCHNEIDER-VOSS**
*Campus Vienna Biocenter 6/1, Rennweg 95B, A-1030 Wien*
*schneider-voss@dialog-gentechnik.at*

**Abstract**. The Austrian society dialog<>gentechnik fosters the dialogue between science and the public. For schools, it has developed several modules and has organized a numer of projects.

## Introduction

Current research in the field of life sciences is making rapid progress and is being driven strongly by the so-called genome research. Potential applications open many fascinating perspectives in medicine and agriculture, for industrial processes and the environment. Relevant issues are, for example, stem cell technology, reproduction technology, molecular diagnostics as well as applications using gene technology. These developments may have a strong influence on the life of single individuals as well as on the society as a whole. The public views certain applications very sceptically. To use the potential of life science research, we need an open dialogue between science and society.

Knowledge of the Austrian public about issues in life sciences is very low compared to other European countries (see Eurobarometer survey 2002). The Austrian media are starting to cover relevant aspects only slowly, often in an unreliable and tendentious way.

The young generation especially should be well informed about the basics, the methods and the applications of life sciences to be able to make informed choices in the future. Nevertheless, the issue is neglected in Austrian schools despite its eminent implications. Actually „genetics“ and „molecular genetics“ are integral, if small parts, in the secondary school curricula for the 8. and 12. grade. Only a few students choose biology as an optional subject and may deepen there their knowledge through school projects work. Thus, means have to be found to integrate current research results into school-teaching. Finally, life sciences should be covered in schools because they represent a growing job market which requires highly qualified personal. Thus, today's students represent the rising generation of future scientists.

The independent Austrian society dialog<>gentechnik is a competent information office on life sciences and fosters the dialogue between science and the public. Its activities are funded by several ministries. For the target group schools, i.e. teachers and students, it has developed several teaching modules and has organized a number of projects. These projects are funded as a whole or financed in part by the Austrian Ministry of Science, Education and Culture.

## Ongoing offers for schools

dialog<>gentechnik has a large collection of basic information on life sciences, press clippings and materials from various sources which is available at the webpage: www.dialog-gentechnik.at.

In the special school corner, dialog<>gentechnik presents new opportunities and possible activities for teachers and students, such as lectures and competitions. In addition, a commented list with recommended links to websites of general interest for schools and interactive websites is listed.

dialog<>gentechnik also offers materials that are helpful for teachers (see table). Finally, it organizes experts for lectures in teacher`s training as well as lectures for students.

Materials for teachers:

- a set of fact sheets covering different aspects of gene technology
- information-packages on current topics such as stem cell research and cloning, etc.
- a list of books and materials to be borrowed free of charge
- a commented link list
- a CD-ROM on the exhibition „gene technology pro & contra" (the exhibition was conceptualized by dialog<>gentechnik itself)
- a set of slides for presentations

**Figure 1.** dialog<>gentechnik organizes teacher's training, where methods used in life sciences are presented and can be tested by interested teachers.

As a major service, colleagues at dialog<>gentechnik answer requests by phone directly or by sending out information materials or possibly by referring to scientists in the relevant field. They generally arrange contacts with experts in science, but may also arrange them with representatives, for example, from official institutions, social sciences and interest groups. About one third of all requests come from students and teachers. Thus, this service is widely used for school projects.

A regular email-newsletter includes information on gene technology and life science topics regarding new scientific results, press releases and media reports on politics, environment, medicine, patents and ethics. It is free of charge and has been ordered by many teachers and students. To satisfy the students' interest for science, it is important for them to be very close to scientists and daily lab life. It is therefore an explicit aim of dialog<>gentechnik to directly involve scientists in its efforts as far as possible. Scientists are very busy with their research and need to be sensitized for public needs for information. Thus, the activities described below are designed to motivate scientists for dialogue.

## 1. Several possibilities have been set up to enable students to picture lab life

- Guided tours are organized for students in research laboratories, where scientists explain their research visually, for example by examining beating heart stem cells in the microscope. They also show their technical equipment from the pipette to the DNA-sequencing machine.
- In one-day-courses, individual students or small student groups carry out experiments such as the isolation of DNA from fruit and vegetables and DNA-electrophoresis.
- During short stays in the lab (one week), students accompany scientists and work together with them on specially prepared mini-projects.
- GEN-AU Summer School 2003/2004 (Ministry for Science, Education and Culture, see below)

## 2. School kit gene technology

Simple experiments in gene technology can be carried out at schools with the experimental "School Kit Gene Technology", which was developed by dialog<>gentechnik. Exact instruction manuals have been worked out and the suitable experiments have been chosen. Care was taken that the experiments need no approbation according to the Austrian gene technology law. Equipment and materials are contained in kind of a protected carrying case. A total of 10 kits are distributed across Austria at cooperating University institutes to facilitate transportation to schools. They can be borrowed from dialog<>gentechnik at little cost. dialog<>gentechnic organizes special training seminars in cooperation with the Austrian teacher-training institutes. There, teachers are instructed in the handling of the kit.

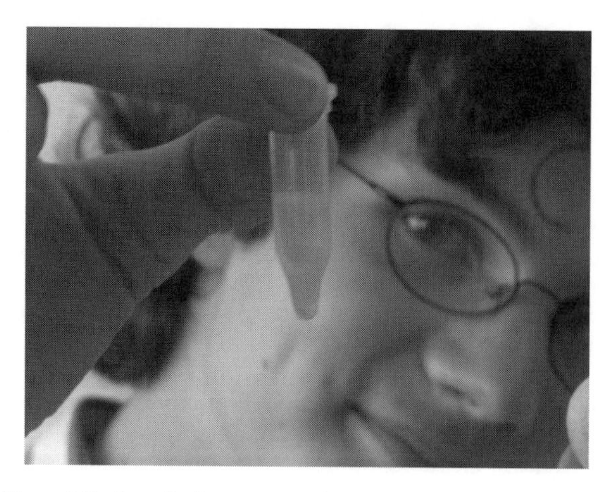

**Figure 2.** Students like it to carry out simple experiments like the isolation of DNA out of fruit and legumes.

## 3. School Competition on „Human Genetics“ 2000/01 (http://www.eduhi.at/humangenetik)

In cooperation with the Austrian Society of Human Genetics, dialog<>gentechnik organised a school competition on Human Genetics supported by the Austrian Ministery of Education, Science and Culture. Pupils from 15 years onwards from Austrian high schools looked into actual research and developments in the life sciences. They acted on their own initiative and in teams. Many varied and creative projects were produced, partly by using the new media. The competition demonstrated that life sciences are extremely suited for project-based and cross-curricular teaching. Defined issues can be dealt with wide interdisciplinarity: English is the scientific language and ethical and social aspects are important for a differentiated view. More than 600 Austrian students participated in about 50 projects, mainly dealing with ethical and social concerns. The entire palette of current human genetics problems was taken up, ranging from new reproduction techniques, therapeutic cloning, the human genome project, gene diagnosis, gene therapy, xenotransplantation to social problems such as handicaps or aging. A total of 18 subjects included not only biology, religion, English and computer science top themes but also the arts and musics. Students also looked for co-operation with representatives from science, economy and politics. Some made surveys and interviews, a few launched social projects. Several projects achieved broad effects. Variety and creativity were mirrored in the means of presentation. Several WEB-pages were made. The competition's format was exported to Switzerland, where a similar project was organized in 2003.

### Student's activities in the scope of GEN-AU (www.gen-au.at)

Currently, dialog<>gentechnik has been commissioned with the public relations for the Austrian genome research program „GEN-AU“, which was initiated by the Austrian Ministry of Science, Education and Culture. It started in 2001 and currently comprises 12 projects in the fields of natural sciences and an additional 6 projects dealing with ethical, social and legal aspects (ELSA). Within the scope of GEN-AU, dialog<>gentechnik conceptualized several activities especially for schools. The activities are carried out in cooperation with the GEN-AU bureau and a communication agency.

## 4. Student's program on the "GEN-AU Day of Discussion 2004"

In 2004, more than 80 students and teachers from technical college and several secondary schools participated at the GEN-AU day of discussion in Graz. Its topic was „Genome research and medicine: what does that mean for me?". Target groups were students and general public. The student's program comprised a morning lecture held by a GEN-AU project leader. The talk was followed by personal meetings and vivid discussions with GEN-AU scientists. This took place in the frame of a poster session, where all GEN-AU projects including the ELSA projects were presented. After the poster session, all students were invited for a final discussion where they asked questions concerning GEN-AU and ELSA projects as well as on other current scientific topics like cloning etc. They were also learning about daily lab life and professional opportunities in this field. Students and teachers were enthusiastic about this kind of dialogue. Finally, several students asked for a place to perform their diploma work. Currently, this day of discussion is being evaluated scientifically by the Stiftung Risiko-Dialog in St. Gallen (Switzerland) to learn about the perception of the chosen format and about the special needs of students. Results will be available in autumn.

**Figure 3.** Personal discussions with scientists are important for students and teachers.

## 5. „GEN-AU Summer School 2003/2004"

In June 2003, the Austrian ministry of science and research took a snap decision to organize a GEN-AU summer school as a pilot project. 17 high school students aged 16 years or more worked for 3-4 weeks in GEN-AU laboratories all over Austria. The students were payed with 200 Euro for their work and an insurance was provided. Students had to write reports which are presented at http://www.gen-au.at. Some of the students wrote newspaper articles about their experience. One article was published in the Austrian daily newspaper „Der Standard" in autumn 2003. Some very motivated students even produced a radio broadcast. After the courses, the coaching scientists and the students themselves answered questions concerning the idea, timing, performance etc. of the summer school. The feedback showed a high motivation at both sides.

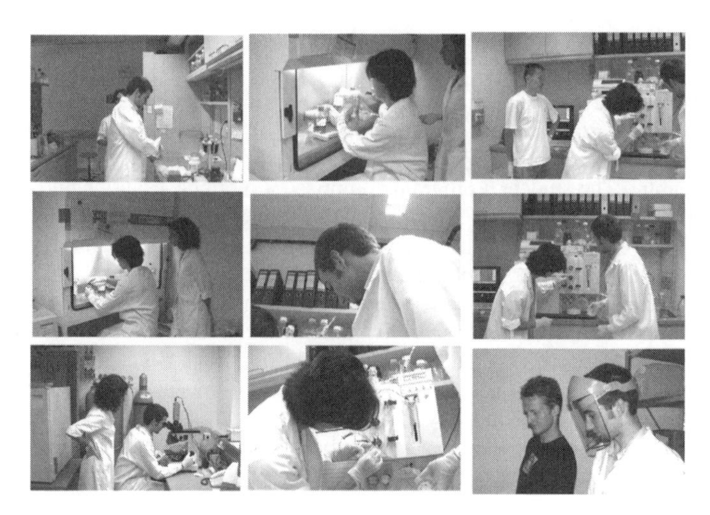

**Figure 4.** The GEN-AU Summer School satisfies the students' interest to be very close to scientists and daily lab-life.

On the basis of the feedback the GEN-AU summer school 2004 was planned on a lager scale. It was announced in February to enable better planning for scientists and students. Again, students were paid with 200 Euro for their work and an insurance was provided. Lab places were organized in the months of March and April and many GEN-AU scientists were interested in the project. At the end, 50 lab places in all but one of the 12 GEN-AU projects were available. Student applicants had to explain their motivation. Applications could easily be distinguished: some students wrote only, that they „would like to get a place", whilst others underlined their special interest with previously performed activities such as participation in the „chemistry olympiade", school project work or the visit of relevant lectures. Outstanding applications even included essays on actual hot topics in the life sciences. Out of 250 applications, around 100 were selected in a first round. Out of these, the scientists could make their choice in personal talks. Again, all students wrote reports which will be presented at http://www.gen-au.at and, in addition, in a year-book. The best documentation will be awarded in an award ceremony in autumn.

The criteria for judgment comprise content, presentation and special qualities such as media work. Feedback from the direct tutors and the age of the students was taken into account. The jury members are representatives of science, school, media and the ministry. Nine prizes and certificates for all participants were provided. As a consequence of the success of the GEN-AU summer school, the Austrian Ministery of Science, Research and Education is currently considering the broadening of the initiative and the involvement of other fields in addition to the life sciences in future summer schools.

## 6. e-Learning project

It is a special concern of dialog<>gentechnik to involve the young generation in the dialogue about actual topics in life sciences. Thus, the society permanently makes efforts to conceive suitable projects and their funding. Currently, it is setting up a national e-Learning project with the aim that students themselves produce cross-curricular teaching material about stem cells and cloning.

## 7. ECOD-BIO network

dialog<>gentechnik is expanding its expertise by coordinating an EU-funded project networking European bioscience information officers. The project http://ecod-bio.org offers a range of useful bioscience informations (frequently asked questions/answers, commented links to excellent websites and teaching materials) as well as a picture pool offering high-quality graphics downloadable free of charge for educational purposes. Within the project, a working group on schools aims at developing a catalogue of recommendations for effective bioscience communication with the target group schools (see http://intranet.ecod-bio.org).

*Science Education: Best Practices of Research Training for Students under 21*
*P. Csermely et al. (Eds.)*
*IOS Press, 2005*


# EMBO Educational Activities

**Andrew MOORE**
*European Molecular Biology Organization,*
*Meyerhofstraße 1, 69117 Heidelberg, Germany*
*andrew.moore@embo.org*

**Abstract.** Education in Europe remains an internationally, and often nationally fragmented activity. In science research, especially, this contrasts with a profession that is highly internationally mobile and that relies heavily on international interchange. Molecular biology, a discipline that is of recognised key importance for our future, is barely covered, even rudimentarily, at the level of practical experiments. Political will and new infrastructures are needed to ensure greater translation of science in the lab to science in the classroom, and to facilitate exchange of diverse educational materials and experience across Europe and the world. Without such international platforms it is hard to see how Europe can improve as a whole.

Do ordinary fruit and vegetables contain genes? Most Europeans probably still think not. Successive Eurobarometer studies in 1996 [1], 1999 [2] and 2002 [3] put the question "Ordinary tomatoes do not contain genes, while genetically modified tomatoes do. True or false?" The respective percentages of correct answers - i.e. "false"- were 35%, 35% and 36%. No change had occurred in 6 years of sometimes frenetic public debate on the very subject of genetic modification technology. "Education" is not the right angle to take when addressing the topic of informing the general public, rather information provision, consultation and dialogue; through these mechanisms, it is theorised, greater public understanding arises. True, it often does. So the question remains as to why a general awareness that genes (and nucleic acids) are present in all living organisms had not grown in the popular consciousness. One could theorise that the work of non-governmental pressure groups and certain discrete worrying events had contributed to a kind of demonisation of DNA via its scheming manipulators (technologists mainly in industry), making it seem almost logical that natural products contained no DNA or genes. However, it transpires that in general the people who did not think that ordinary tomatoes contained genes were also those who were most supportive of GM technology [4]. Then perhaps it was partly a quirk of linguistics in some countries that led GM products to be known as "gene-food" (gene-tomatoes etc.). Whatever the reasons, the message is clear: imparting scientific literacy and the tools to analyse topics of relevance to our lives is most easily done early in life (in the relatively uncomplicated environment of the school); later things can become much more difficult, people fall into so-called “hard to reach” groups, and the task of providing objective information against a tide of uncontrolled and un-peer-reviewed information becomes fraught with problems.

Scientific literacy may be defined as the capacity to use scientific knowledge to identify questions and to draw evidence-based conclusions in order to understand and help make decisions about the natural world and the changes made to it through human activity [5]. In testing this practical ability, rather than more academic indicators of school education, the PISA study [5] (Programme for International Student Assessment), co-ordinated by the Organization for Economic and Co-operation and Development is the most important so far. Furthermore, it tested the degree to which students developed the skills to become life-long learners. In an ever faster changing world, this is particularly important. Though one may not have learnt a subject at school, the ability to get up to speed when it hits the headlines is a must if one wants to play a constructive role in the public debate.

But understanding science also includes possessing a grasp of how scientific progress is made, the nature of scientific results (especially in non-precise subjects such as biology), the time scale involved in the emergence of applications from research discoveries, and essentially what doing research at the bench is all about. This is where the link between school teachers and research institutions is crucial. The school laboratory is a special place where young people get their first taste for the excitement of a scientific experiment; it must not be left to moulder in a quagmire of purely descriptive science and experiments that demonstrate classical knowledge in recipe style. Increasingly, however, other sources of educational material and practical experiences in molecular biology are threatening to take the focus away from the school laboratory, threatening its important role. This also neglects the creativeness of teachers, and the many good resources that they and scientists have developed for the school laboratory. What is currently lacking in most European education systems is an infrastructure in which the dissemination of such resources can take place, and in which the teaching and scientific support necessary to build the practical skills and confidence among teachers are provided. And what is lacking above that is a mechanism for generating universally higher standards in science (not just biology) education across Europe by integrating the diversity of good materials and best practise that we have scattered across our continent.

Though the precise results of the first PISA study, published in 2001, may be a dim memory for most, or a dull embarrassment for some, the general picture of fragmentation and widely varying standards among European countries remains as a striking feature of our educational systems. In science research, we observe that, taking all disciplines together, a country’s gross domestic product (GDP) and GDP per capita correlate extremely well with scientific output [6]. However, there is no such obvious link between a country’s wealth and its standard of education [5]. In Europe, fragmented as we remain at many levels, this is our greatest hope for building the foundations of a scientifically literate, critical and engaged population, and for producing the talented minds of the future. It means that whatever a country’s misfortune in economic terms (much of which is often historical) its potential to produce highly educated citizens is not primarily an economic matter. Furthermore, viewed as an evolving organism, Europe has an envious degree of diversity of ideas and concepts resulting from its cultural and linguistic diversity. Via exchange of such ideas and experiences it should be able to profit from the best, hence increasing its classical fitness. At the same time, it must be careful to preserve diversity at all levels of education; most importantly at the level of the individual teacher, who must be given the freedom to develop continuously throughout his/her career, introduce new ideas into the classroom, and in the natural sciences more truly experimental work. To harness such diversity rather than suffering from its drawbacks, however, it is necessary to convene the "movers and shakers" in the national systems at regular intervals at international level.

What can be said of the diversity in Europe, can also be applied to global diversity. Here again, PISA has set some interesting pointers for further developments and exchanges of experiences. Scientific research these days is an increasingly global venture depending on international collaborations, exchanges and centres of excellence. Very few laboratories on the planet do not profit from international exchange. At present, very few school *do.*

The European Molecular Biology Organization (EMBO) has established a successful platform for training and exchange among teachers from across Europe (and beyond). The annual international practical workshop for teachers accommodates 120 teachers, typically from over 20 countries, for 2 days in Heidelberg, Germany. At the workshop, they have the chance to do hands-on practical experiments in molecular biology, browse an exhibition of some of Europe's best teaching resources, visit the research labs of the European Molecular Biology Laboratory, and hear talks from leading scientists. The workshop model was extended in 2003 and 2004 via an EC grant to 8 other locations in Europe and Israel, hence further expanding the network of teachers who are benefiting from international exchange. These workshops form part of the larger project co-ordinate by EMBO, "Continuing Education for European Biology Teachers" CEEBT), in which the European Molecular Biology Laboratory plays a major role. Oversubscription of the Heidelberg workshop in particular points to a growing realisation that international exchange is seen as important by biology teachers. EMBO gives access to teaching materials from these workshops via its European Network for Biology Education web page [7], which hosts a database of good teaching resources in biology worldwide. It has also produced a DVD mini-documentary and guide to help others organise a similar event.

Dissemination of the best materials and experience is a key to raising standards uniformly in Europe, while preserving the diversity of Europe. But it is also important to have knowledge of national systems and their constraints. EMBO gathers this information by interacting with teachers in its network, and is able to bridge the gap between school and research through scientists in the EMBO membership (around 1.300 in 25 member states), EMBO fellowship holders and Young Investigators, and many other scientists. In our experience, the interaction of teachers and scientists can not be left to chance. It may seem logical that they should interact, and one might expect interactions to result as a natural consequence of meeting at a teachers workshop. However, there are considerable "activation energy" barriers on both sides. Scientists in general do not understand well the context in which teachers work, what they can and cannot do, and how a laboratory experiment needs to be adapted in order to be performed in school. Teachers may not have the time or confidence to visit a scientist, and may not know where to start in terms of asking him or her for help.

Likewise, our experience shows that apparatus pools, if established from decommissioned items by research institutes, do not suddenly take off and circulate happily among teachers. Major legal constraints and resulting inhibitions largely prevent this, despite the fact that the financial realities of most schools make it impossible to buy brand new equipment. This results in fewer than 30% of the schools in Europe possessing even the simplest apparatus to demonstrate rudimentary molecular biology. Uncertainty or lack of confidence among teachers (especially older ones) that they can actually use and maintain the equipment properly compounds the problem. Researchers must reach out to teachers, and help them build the confidence they need to interact more with research institutes and departments, and perform simple experiments in their classes.

However, it is not as if these measures on their own would improve the teaching of practical science in schools. Objections to the logic that more practical work in molecular biology must be incorporated into school timetables are: that not enough time is available given that biology is just one among many subjects, and that timetables only allow short slots for practical work. Clearly, many simple experiments can be done in a short time period, but more interesting ones require project based practicals, a luxury that few schools in Europe, notably the European and International Schools, have. More project based learning opportunities are needed in all the sciences if students are to learn the method and way of thinking of science, rather than learning it as cookery-style recipes.

All teachers at EMBO international workshops would like to do more practical experiments and projects in "real" molecular biology. Experiencing practical molecular biology cannot be left to one-off visits to research institutes; it must become an integral part of teaching. Workshops in molecular biology are clearly necessary and in demand in greater numbers across Europe. Introducing the latest research via talks from the researchers themselves, hands-on practical experiments, and an exhibition of diverse teaching resources all give teachers the tools and material they need to inspire pupils. Furthermore, the great creativity of many teachers should be given the dimensions of resources and time in which to work in order to prevent the wasted opportunities that currently mar the current education systems.

A problem facing all biology teachers regardless of country is a lack of motivation on the part of the pupils. This is the kiss of death, for without interest, learning is a temporary and forced exercise associated with distant school days. Teaching is certainly a dynamic and stimulating profession, because it essentially combines learning with the creative act of communication. At EMBO workshops, as at many others, it is evident that for highly-motivated individuals it is a rewarding challenge, a profession that deserves to continue attract the best, and a profession that must be supported properly at international level by the scientific community and policy makers. After all, our joint futures are shaped by an increasingly internationally mobile consumer society and an already highly internationally mobile pool of researchers. Addressing education at international level, therefore, is not for the future; it is for now.

**References:**

[1] Eurobarometer 44.3 OVR 1996, European Commission, Brussels
[2] Eurobarometer 52.1 OVR 1999, European Commission, Brussels
[3] Eurobarometer 58.0 OVR 2002, European Commission, Brussels
[4] GM Nation? The public debate: http://www.gmnation.org.uk
[5] Knowledge and Skills for Life; First results from PISA 2000, OECD (and see also http://www.pisa.oecd.org/)
[6] Science Citation Index (Institute of Scientific Information) and The World Fact Book (US Central Intelligence Agency)
[7] http://www.embo.org/projects/scisoc/education.html

*P. Csermely et al. (Eds.)*
*IOS Press, 2005*


# Extending the German Pupils Academy to Younger Secondary School Pupils: The German Junior Academies

**Harald WAGNER**
*Bildung und Begabung e.V.,*
*Godesberger Allee 90, 53175 Bonn, Germany*
*wagner@bildung-und-begabung.de*

**Abstract.** The "Deutsche SchülerAkademie" (German Pupils Academy) is the most significant outside-of-school residential programme for gifted senior secondary school pupils in Germany. Following its description at the first NATO / UNESCO research workshop an extension of the concept to $7^{th}$ to $9^{th}$ graders is presented here. Results from two regional academies during a three year pilot phase have shown that the concept for the "seniors" with minor adaptations can successfully be applied to the "juniors". In cooperation with other organizers of similar programmes the label "Deutsche JuniorAkademien" (German Junior Academies) has been created to promote the idea of high quality residential summer academies all over Germany.

## Introduction

In the last paragraph of my contribution to the 2002 Research Workshop at Visegrad [1] I mentioned plans for future developments of the "Deutsche SchülerAkademie" (German Pupils Academy), a residential summer programme for highly able and motivated senior high school pupils (grades 10-12, 16 to 18 year-olds) in Germany: 1) an intensified cooperation with our Eastern neighbours, 2) the establishment of a summer academy for junior high school pupils with a strong emphasis on mathematics and science. As it was possible to raise additional funds both plans could be materialized.

A grant from the Haniel Foundation enabled us (that is Bildung und Begabung e.V., a non-profit association financed mainly by the German Federal Government and by the Stifterverband für die Deutsche Wissenschaft [Donors' Association for the Promotion of Sciences and Humanities in Germany]) to establish a multi-national 16 days summer academy with four courses and a total of 64 participants from Germany, Poland, Hungary and from the Czech and Slovak Republics. The working language is German; the venue is a boarding school in the Benedictine abbey of Metten, Bavaria. The multi-national academy follows the same proven educational and organizational principles as the regular academies. During a pilot phase from 2003 to 2005 three such academies will be held. After the first two very successful programmes there are good chances for a continued and even extended support by the Haniel Foundation after 2005. A donation from the Zurich Group (an international insurance and financial services company) provided the necessary financial support for the realization of the other plan, which is the focus of this paper.

## 1. The need for early talent development

While our summer academies for senior secondary pupils undoubtedly play a most important role for the development of talents, motivation, study habits, self-esteem, social contacts etc. (cf. [2]), it was felt from the very beginning that this type of programme should already be offered to younger pupils as is practised by e.g. the Center for Talented Youth (CTY) at the Johns Hopkins University, Baltimore, or the Talent Identification Program (TIP) at the Duke University, Durham, N.C.

In Germany gifted and talented pupils in the lower level (grades 5 – 10) of secondary schools face little additional challenge or support from extracurricular enrichment programmes. They all have to follow the prescribed timetable of their class without the possibility to select subjects. Therefore, an early encounter with interesting and demanding topics in outside-of-school settings is highly desirable.

A matter of serious concern in Germany, as in many other countries, is the dramatically declining number of university students in physics, mathematics, chemistry and engineering. Many pupils tend to avoid these presumably "tough" subjects already in the upper level of secondary school (grades 11 – 13) thus excluding these subjects more or less as options for university studies. Therefore pupils should have an opportunity much earlier in their school years to explore their interests and abilities and be encouraged to engage in activities in these fields – and this is especially important for girls who are still severely under-represented and who supposedly form a largely untapped reservoir of talents.

## 2. Basic considerations for a junior academy

By the end of 2002, when it came to the planning of a first residential summer academy for the lower secondary school level, it was decided to aim at pupils in grades 7 and 8 (12 to 14 year-olds) and to put a strong emphasis on mathematics and sciences without, however, totally neglecting the humanities. Furthermore it was decided to restrict participation to pupils of the federal state of Rhineland Palatinate where the academy was to take place in a boarding school in Neuerburg, a small town near the border to Luxembourg. So it was possible to successfully apply for an additional grant from the State Ministry of Education which helped to reduce the participation fee for the pupils by 30 percent (from 500 to 350 Euro).

Applicants were expected to be highly motivated and to have already demonstrated exceptional achievement in or out of school, e.g. successful participation in an intellectually challenging competition. The majority of applicants came through recommendations from schools where each school could nominate not more than one or two candidates.

The number of courses was reduced from six in the "senior" academies to four with up to 16 participants each. This should help reduce possible problems during the three years pilot phase.

## 3. Junior academy 2003

The Junior Academy Rhineland Palatinate was held from 24 July to 9 August with a total of 62 boys and girls. The academy was opened in the presence of the participants, their parents and the staff by the Rhineland Palatinate State Minister of Education and a board member of the Zurich Group. On the final day the participants proudly presented the methods and results of their course work to their parents. The following courses were offered:

1. Planes, knots and graphs (an introduction to graph theory)
2. Evolution – what's behind it? (molecular biology and evolution)
3. Applied aerostatics and aerodynamics (the physics of flying)
4. Die Nibelungen (Middle High German epic poem from the 13$^{th}$ century which has influenced Germanic arts like no other literary work over 800 years; introduction to literary science and the reception of the work)

Course work covered about half the time of the academic days while the other half was filled with all kinds of activities where pupils from all courses would mix – music and sports being the most important ones. About 60 percent of the pupils played an instrument, therefore a full time musician was part of the academic staff (site director and assistant, and eight instructors, two for each course) as is the rule in the "senior" academies. He conducted daily choir rehearsals and arranged and advised instrumental ensembles to work towards a quite remarkable public concert. Musical activities have a strong integrating effect on the whole academy. Other activities were drama, visual arts, games and excursions.

In addition to this academy, two other regional programmes were offered in 2003 by institutions in Saarland (30 participants) and Baden-Württemberg (60 participants) with a duration of 8 and 14 days. They aimed at pupils in grades 7 to 9 and 8 to 9 respectively.

## 4. Results

On the whole this first Junior academy was a big success. The participants were enthusiastic about the 16 days. They praised the excellent working atmosphere, the amount of independence and responsibility they were granted by their "cool" instructors, and the absence of "school marks". The course work was judged to be demanding, challenging, and rewarding. The instructors were very pleased with the high motivation and the quality of course work and presentations given by the pupils. They observed, however, certain differences between the "juniors" and the "seniors" in the traditional academies:

1. Many of the juniors had difficulties systematically developing and following a path from a specific problem through to a solution. They lacked not only methodical abilities but also endurance and the insight to be responsible for their own learning.
2. The pupils needed many impulses from outside to achieve learning results, they needed detailed instructions and task descriptions but then they set off with joy, dedication and creativity, so that the instructors had to take care that the pupils did not become overexcited.

3. These pupils usually hold top positions in their home school. Thus they are not used to criticism and can hardly handle it. They had to learn to accept critique as a chance for improvement and growth.
4. The "juniors" seemed to be less willing and able to devise their own leisure time activities. Due to their younger age and less practice their musical abilities (playing an instrument) were less developed; this influenced the selection of scores for concerts.
5. While most of the concept and structure of the "senior" academies can be transferred to the junior academy without substantial changes some aspects need special attention: the juniors need more guidance and an emphasis on the development of autonomous learning and knowledge acquisition and the ability to work in teams.

To improve the group dynamics from the very beginning and to get detailed information about the individual abilities and knowledge levels of the participants it was recommended that a preparatory meeting with the pupils be held well ahead of the next academy in the following year.

## 5. Junior academy 2004

During the summer of 2004 Bildung und Begabung e.V. organized a second junior academy at the same site in Neuerburg (more details may be found in the internet under www.schuelerakademie.de/dja/2004/programm.html). The size was kept the same as in the year before (four courses, 64 participants) and the duration was again 16 days. This time financial support came (once again) from the State Ministry of Education and from the Klaus Tschira Foundation, Heidelberg while the participants paid a nominal fee of 350 Euro which could be reduced or waived for families in need. The participants and the other two partners each provided roughly one third of the total budget of 63,600 Euro. The following courses were offered:

1. Discreet communication (encoding and cryptography)
2. Ecology of running waters (examining the properties of the ecosystem of a river)
3. The experience of the starry sky in history and today (astronomy and culture)
4. On the road to Canossa (introduction to historical research)

Three months prior to the academy a preparatory meeting of the instructors and participants was held on a weekend from Friday to Sunday which helped them considerably to become familiar with each other and with the boarding school, to clarify expectations with regard to course work, and to learn and improve techniques of knowledge acquisition and presentation. As a result the academy itself started much smoother and faster and allowed productive work from the very beginning. On the whole, the second junior academy was very successful and we again received enthusiastic comments from the participants.

The following letter from Michael (aged 13) may stand as an example for numerous similar responses:

> *"It was great to meet with other youngsters of the same age and to work with them on extra-curricular subjects. The instructors explained everything very well so everybody could understand. It was a pleasure to meet other 'gifted people' although I hate this expression because neither I nor other participants in the academy want to be regarded as such. It was a joy to find contact with other especially motivated and able youngsters and to talk to them about complicated*

*subjects without at once being called 'nerd' or 'braggart'. So far I have never been able to talk so intensely about all aspects of life. There I received honest, not somehow memorized replies and advice to many questions. We could exchange our experiences, positive as well as negative ones, regarding school, for instance. Quite often we sat together in the evenings in a room or outside talking to each other and having a lot of fun. I liked the fact that course work proceeded without the usual school stress although we learned more and much faster than in school. This was due also to the motivated participants who, quite different from school, followed the course work intently. Because of all these very nice experiences, parting from the academy was especially difficult for all of us and therefore we vented our emotions, crying for hours. I very much hope that we all stay in contact with each other and with the academy because this time is and will remain an incomparably wonderful event in my life."* (Translated by the author)

## 6. Additional junior academies

In Darmstadt (Hesse) an academy was held by the Kinder- und Jugendakademie Südhessen e.V.; it lasted 14 days with four courses and 15 participants each from grades 8 to 10 (www.kijash.de). A preparatory meeting was held six weeks prior to the academy. The course titles were

1. Memory and design (art, architecture, psychology)
2. Can luck be calculated? (probability theory, set theory, combinatorics)
3. "Attention, top secret!" (introduction to cryptography)
4. "It's your world" (introduction to the function and structure of the United Nations)

In Homburg (Saarland) an academy was organized by Beratungsstelle Hochbegabung Dillingen which lasted 10 days (www.iq-xxl.de). The 50 participants came from grades 7 to 10. This academy was somewhat differently organized as it offered several workshops which referred to the common theme "communication". Participants attended each workshop in groups of about ten for some days before moving to the next workshop. By this they received an overview of the general theme while practicing different methods of knowledge production. Parallel to the workshops methodical training was offered on research methods, observation and documentation, visualization and presentation techniques. The workshops centred around the following themes:

1. Communication in the animal kingdom
2. Communication technology: Natural techniques of communication, analysis and technical realization
3. Rap, pop, rock, chanson: Modern French music
4. Body language

In Adelsheim (Baden-Württemberg) the Oberschulamt Karlsruhe organized an academy which lasted 13 days with six courses and 15 participants each from grades 8 and 9 (www.scienceacademy.de). A preparatory meeting was held 14 weeks prior to the academy and a reunion took place six weeks after the academy where a documentation of the course work was produced. The course titles were:

1. Biological engineering? Molecular biological experiments and their ethic reflection
2. Pinball, breakout & friends (physics of movement)
3. Trendy, but somehow strange: Alcopops & Co. (chemistry of food and drugs)
4. How robots tick (robotics)
5. A look at nature: Examining woods and waters
6. Applied research/development/economics. A project with a serious background

**Figure 1.** The logo of the German Junior Academies.

All of these programmes were offered under the umbrella of "Deutsche JuniorAka-demien" (German Junior Academies). This label shall stand as a brand for high quality residential summer programmes for highly able and motivated pupils in the middle level of high school in Germany and shall inspire other institutions and federal states to develop similar programmes. To be included in the Deutsche JuniorAkademien a programme has to fulfil certain requirements concerning minimum duration, qualification criteria for participants and instructors, the variety of disciplines presented in the courses and the provision to allow for reductions of the participation fee for needy families. For the following years junior academies are planned in cooperation with these and additional partners, Bildung und Begabung e.V. playing the role of coordinator of these efforts.

**References:**

[1] H. Wagner, Talent Development in Residential Summer Programmes. In: P. Csermely and L. Lederman (eds.), Science Education. Talent Recruitment and Public Understanding. IOS Press, Amsterdam, 2003, pp. 117-128.

[2] H. Neber and K.A. Heller, Evaluation of a Summer-School Program for Highly Gifted Secondary-School Students: The German Pupils Academy, *European Journal of Psychological Assessment* **18** (2002) 214-228.

*Science Education: Best Practices of Research Training for Students under 21*
*P. Csermely et al. (Eds.)*
*IOS Press, 2005*



# ....You Do and You Understand: Hands-on Experiments at the XLAB

**Eva-Maria NEHER**
*XLAB-Goettingen Experimentallaboratory,*
*Goldschmidtstrasse 1, 37077 Goettingen, Germany*
*emneher@xlab-goettingen.de*

**Abstract.** XLAB is an educational institution, which wants to bridge the gap between high school and university. XLAB organizes experimental courses in Biology, Chemistry, Informatics, and Physics for classes and individual students from EU-countries and from all over the world. The students do intensive experimental work with state-of-the-art-equipment. Theoretical teaching by experienced scientists runs parallel with the experiments.

## Introduction

In nearly all industrialized countries the number of students enrolling in natural science studies at universities has been decreasing dramatically for more than 15 years. On the other hand science and technology provide the key to the problems and challenges that our societies are facing today.

Much effort has to be invested to encourage young people to pursue scientific careers. Young people have to get enthusiastic about the great research adventure of today. Students should get to know how to do research: what it means to work in a laboratory, what it means to solve a theoretical problem, and for what purpose a computer is really needed, instead for fun. That means students should get to know the reality!

Of course it is not enough to have an open-doors research institute for a day; it is not sufficient to give a public lecture, and it is not enough to visit students at school and to show some spectacular experiments.

There is a tremendous need of changing the teaching methods at schools, both at primary and secondary schools. This seems to be same in all industrialist countries as it was shown in the results of the worldwide investigations like TIMSS and Pisa.

Of course for primary schools and secondary schools different teaching methods are necessary. Education in primary school will focus mainly on the phenomena observed in the experiments whereas in secondary school the topics should reflect more and more the latest results in research. In both cases hands-on experiments play an important role and theory should come along with the experiments and not separately or exclusively, as this is the reality in most school systems.

The economical situations of high schools do not allow to install sophisticated experiments: the equipment is much too expensive and teachers are not trained in supervising experiments on a more or less scientific level.

In contrast, the establishment of central laboratories makes good economic sense: Central laboratories can serve regional schools and may also be accessible nationwide and - as is the case for XLAB – worldwide.

## Aims of XLAB

Concurrent with the Bologna Process XLAB is following the general aims of the EU in promoting the attractiveness of the European Higher Education Area and promoting the mobility of the students and encouraging them to take up university studies abroad. In particular XLAB tries to raise student's interest in science subjects in order to increase the number of future scientists.

## Teaching at the XLAB

The XLAB tries to provide an atmosphere of real research laboratories with authentic tools and machines and most important lecturers, who are experienced scientists.

XLAB offers a variety of practical experiments in biology, chemistry, geosciences, computer science, mathematics, and physics. The experiments are designed and supervised by scientists. Scientists and science schoolteachers work together in a very tight collaboration; the performance of the experimental courses is supported by qualified technical assistance. This guarantees the specialist scientific knowledge, experienced didactical teaching, and successful performance regarding the technical prerequisites.

Students work in the laboratories for the entire day. They concentrate on one subject; that means there is no interruption by other lessons as it is the case at school. This provides an intensive learning at a level, which can be compared with university teaching.

Most of the students are very satisfied with their progress in learning by doing hands-on experiments. Others learn, that taking up university studies in science would be not the right decision for them. This is very important and prevents them from frustrating experiences during later university studies. Another group of students get to know about non-scientific-careers in the field of science, and technology, which is also very important, since well-educated technical assistants are of great demand in scientific research.

## Target groups

The target groups are:

1. School classes with their teachers coming from Germany and all EU-countries. Classes stay for one to five days after having made special appointments.

2. High school and 1st year university students coming individually, attending weekly courses. Students coming from all countries are heartily welcome! In contrast to the International Science Camp, there is no limitation to the number of participants coming from one country.

3. International students participating in the XLAB Science Camp for three weeks during the summer holidays. The number of students representing one nationality is limited to 2-3 only.

## Worldwide acceptance of XLAB's offer

XLAB started in August 2000 and after three school years more than 10 000 students did experiments at about 22 500 days (students x days). Many individual students and teachers with their classes come again and again. (Fig.1 and 2)

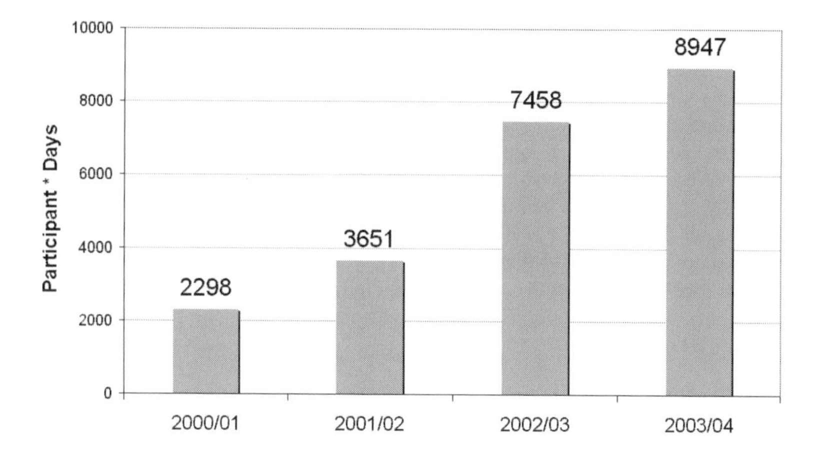


**Figure 1.** Number of students since august 2000. The number students in 22003/04 conforms 25% of the capacity XLAB will have when mowed to the new building.

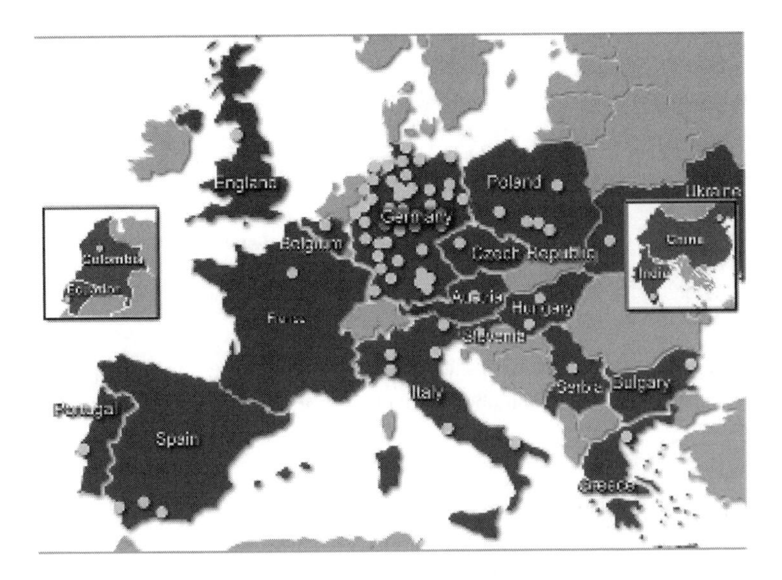


**Figure 2.** Students come from all over the world to attend the International Science Camps. But international students and classes are also welcome to participate in the weekly courses all year round.

In November 2004 XLAB will move to a new building, especially designed for teaching high school students in the various science subjects. The four story – nicely colored -building has laboratories and seminar rooms to teach 160 students simultaneously. Each floor is dedicated to one of the four subjects but nevertheless the interdisciplinary character of modern research can be experienced. For example, all the equipment needed for studying neurobiology will be found in the physics department and experiments in biochemistry will be carried out in the chemistry laboratories. Moving to the new building does not mean, that XLAB will become independent from the research laboratories. Whenever special measurements like NMR-Spectroscopy, electron microscopy and many others are necessary, the students work within the different research laboratories of the university and the Max-Planck-Institutes in Göttingen.

## International Science Camps

In 2003 and 2004 XLAB organized International Science Camps. 2003 the camp brought together 44 students from 12 different countries and in 2004 we count 35 students from 17 different countries. Teaching is done in English and, of course, the common language of the camp is English as well.

Science camps last for 3 and a half weeks. The scientific program takes three weeks. Each student chooses three weekly courses out of a program of 12 different experimental courses in the fields of natural sciences. (Fig. 3) Students present the results of the each course to each other on Saturday afternoon. In the third week an excursion to Berlin highlights the social program.

**2nd International Science Camp 2004**

| 1st week | 2nd week | 3rd week |
|---|---|---|
| 16th - 20th August | 23rd - 27th August | 30th August - 3rd September |
| A1<br>Neurophysiology | B1<br>Molecular Biology | C1<br>Developmental Biology<br>& Embryology |
| A2<br>Analytical Chemistry | B2<br>Organic Chemistry | C2<br>Biochemistry |
| A3<br>Physics of Flying | B3<br>Laser Physics | C3<br>Astrophysics &<br>Astronomy |
| A4<br>Informatics - Optimisation | B4<br>Anatomy | C4<br>Material Physics |

**Figure 3.** The courses of the 2$^{nd}$ International Science Camp.

Science is international, and the scientific community resembles a worldwide family. There are no prejudices with respect to nationality and political or religious affiliations. Scientists from all over the globe, having a common interest in special research topics, meet each other at international congresses and workshops at various places on all continents.

The XLAB international Science Camp conveys this experience to our future scientists. Young people, 17-21 years of age, will regularly meet in summer in Goettingen, discover their common interest in sciences, work and live together, make international friendship across the borders of cultural heritage, get to know the country of their host, and get a feeling of what it means to become a member of the various international scientific communities.

*Science Education: Best Practices of Research Training for Students under 21*
*P. Csermely et al. (Eds.)*
*IOS Press, 2005*


# Giftedness and Thinking

**Ida FLEISS**
*An der Bastion 1 A, D-50679 Köln, Germany*
*mittring.fleiss@tiscali.de*

**Abstract.** In this report I want to point out some special cognitive abilities which are found in gifted persons, especially in inventors and some scientists. How does an inventor come to new ideas? How does a scientist find the solution for a complicated problem? Can gifted children learn specific techniques for inventions? Samples [1] suggested ways to help people form metaphors to train creativity. He noted that almost all theory-making in science is metaphoric. In addition to studies in the relevant research papers we have interviewed a number of successful inventors (chemists, technician, managers) in order to find out, how they are thinking when they handle matters which grow into inventions. Their answers and comments confirm the theories and findings of Rothenberg [2,3], Mittring [4], Sternberg [5,6]. In the following I will describe and comment these three postulated cognitive abilities and ways of thinking.

### "Janusian thinking"

This kind of thinking is named after the two-faced Roman god Janus, who could look into two opposite directions. People with this ability can actively conceive of two opposites simultaneously and thereby process two disparate views on a subject in parallel. This type of thinking is believed to be a key step in the process of the creation of the kind of scientific theories and/or discoveries of such people as Einstein, Darwin, Watson, Pasteur and Fermi. This ability of thinking also might be essential to the creative thinking processes of literary critics, poets and philosophers [2,3].

One inventor who possesses over 50 patents for chemical inventions told me some examples how he is thinking: he seeks opposites like "hard weakness", "heavy and light at the same time" or "hot and cold" a.s.o. and thinks how he could overcome the opposites or how to use one opposite to fit into the other. The way of thinking in oxymorons (opposites) is very enlightening and offers lots of insights.

There are ways to measure Janusian thinking. An offshot of the Kent-Rosanoff test is used to measure this kind of cognitive ability. This test involves having subjects respond to ninety-nine orally presented words with the first word that comes into their heads. Responses to these very common words are then divided into 3 categories:

1. response words that mean the opposite of the stimulus word
2. "primary" nonopposites responses (i. e. the most common answer)
3. all other responses.

Among others this test was applied by **Rothenberg** [3] to nine Nobel laureates in science, spanning physics, medicine and chemistry. As expected, exceptionally high scores in Janusian thinking were reported among the Nobel laureates. They answered with an opposite more than 60% as compared with student groups. Moreover, as the creativity level of respondents rose, there was a significantly faster response time associated with Janusian thinking answers.

My personal suggestion is to introduce this concept of Janusian thinking in the curriculum for teaching gifted children with strong interests in science, technology and mathematics.

## "Metaphorical intelligence"

The greek meaning of metaphor (meta pherein) is "to transfer, to carry over". That means, that two originally separated meanings produce a new meaning when combined, the metaphor. For instance "bottleneck" The two words have different meanings when separated: a bottle as a subject and a neck as an anatomical entity. Together they express something like a narrow pass, a defile [4].

Every language has its own metaphors. They cannot be simply translated word by word. They have to be decoded and then encoded again. This process requires a special ability, the metaphorical intelligence.

There are situations in which metaphors are used to mask or disguise facts, to dissemble things, to hide a specific meaning or to wip out or to distort things: in diplomacy, in difficult political negotiations or in espionage a.s.o. Those persons are lost who cannot grasp the true meaning of a metaphor, or to discover fallacies.

Very important is metaphorical intelligence for translators and interpreters at conferences. Literally translation of idioms or metaphors can lead to confusion, misunderstanding or even to confrontations.

Also in daily life we can find situations in which one person catches the true meaning of a conversation, the other does not. They simply cannot "read between the lines", as we German call it.

The measurement of metaphorical intelligence can be done by tests which contain proverbs and metaphors which should be matched. Or tests in which metaphors should be explained.

According to **Mittring** [4] it is a special gift to understand what others mean by using a specific metaphor. He calls it "metaphoric intelligence". He describes it as a decoding ability, an ability to understand complex contextual structures and to see or listen beyond given structures. He postulates the following steps:

1. **perception,** realization that a given text is a metaphor,
2. **disintegration,** encoding the elements of the given metaphor from the literal meaning to its sense,
3. **integration**, embedding the sense of the metaphor into the context,
4. thus **understanding** the whole text in its metaphorical meaning.

Mittring has worked out a typology of metaphorical intelligence. He shows hat this form is distinctive from social and emotional intelligence.

He also suggests to include metaphors in curricula for natural sciences, mathematics and informatics, thus stimulating gifted children to develop their metaphorical intelligence.

**"Analytical insight"**

The ability to generate insights is considered to be central to intellectual giftedness. This ability is the common denominator of diverse talents like for instance mathematical inventiveness or extraordinary skills in natural sciences [5].

Sternberg and Davidson have found out, that there are three different kinds of insight. These three kinds form what he calls the "creative intelligence" (cited from Storfer [6] p. 374):

- **selective encoding** - or the ability to discern what pieces of information are relevant to the solution of a problem (from among considerable irrelevant information) and to discern in what way(s) they are relevant,
- **selective combination** - or the ability to discern how to combine what might originally be seen as isolated (or at least not obviously related) pieces of information into a united whole,
- **selective comparison** - or the ability to compare newly acquired information to that acquired in the past and to discover unexpected or unusual relationships between the new and the old information.

These authors believe that it is difficult to isolate "creative insight" because this kind of insight involves the ability to successfully employ all three of these separate but related processes rather than depending on only one of them.

Sternberg and Davidson also [6] have shown that gifted people do possess a greater ability to generate insights in all three of these areas, in that they need less information to solve insight problems that average gifted people who need much more information.

The ability to generate analytical insight enables us to solve problems, but what makes us to seek new solutions?

**Curiosity** leads us to seek new problems. Berg and Sternberg [5] found out that there are differences in the interest for new things and the competence to handle new things adequately between average gifted and highly gifted children. Gifted children are more curious than not gifted ones.

**References:**

[1] Samples, B. (1976): The metaphoric mind. Reading, MA: Addison-Wesley
[2] Rothenberg, A. (1979): Einstein's Creative Thinking and the General Theory of Relativity: A Documented Report. American Journal of Psychiatry, 136, pp.38-43
[3] Rothenberg, A. (1982): Janusian Thinking and Nobel Prize Laureates. American Journal of Psychiatry, 139, p.122
[4] Mittring,G. (2004): Die metaphorische Intelligenz. In: Diskussion Nr. 315
[5] Berg, C. A. & Sternberg, R. J. (1985): Response to novelty: Continuity versus discontinuity in the developmental course of intelligence. Advances in Child Development and Behavior, 19, p. 1-47
[6] Storfer, M. D. (1990): Intelligence and Giftedness. The Contributions of Heredity and Early Environment. Jossey-Bass Inc., Publishers California

# Field-work Experience as a Research Initiation for Students: The Case of Applied Geophysics

**Michael S. ARVANITIS**
*Euroscience Greek Regional Section,*
*PO Box 3125, 10210, Athens, Greece*
*mike@georasis.gr*

**Abstract**. Field-work can become a research initiation for students and it can make them put in real the theory they have learnt in class. Not all lessons can offer this option and here we describe the experience of field-work for the course of applied geophysics.

## Introduction

It's very rare for a class to have lessons in an open-area, meeting and facing real-time conditions, participating actively in what is called the making of scientific knowledge. Usually, children and teachers are limited to what a classroom can offer. Despite the technological progress (i.e. virtual reality) and its ability to simulate real conditions, a classroom falls short of field work. Geophysics, on the other hand, is a science based on experiment and on out-of-school activities. If planned in a smart way, field-work can become the spark which will light the fire of curiosity to children; in other words it has the potential to become the basis of a future researcher.

## The principles of field geophysics

Geophysics is the science which focuses on the earth's interior and its internal procedures. The most difficult part of a geophysical course for students is not theory, which is closely related to wave physics and mathematics, but to give students a palpable example of how the theory can be applied to a real project. It's different, i.e. saying that different velocities of seismic waves exist, a difference which depends on the depth of the earth, than using these velocity-variations on a real project in order to discover a gas or an oil deposit.

Identifying the teaching sites is probably the most important step of the whole procedure. In order to keep students motivated, it is necessary to plan field-works at sites of real interest; these sites can be linked closely to the student's interests and can be interdisciplinary, i.e. a combination of geophysics and archaeometry or arhaeology in order to discover burden antiquities without excavating, or a combination of geophysics and environmental studies in order to find an aquifer or sewage.

The choice is not always as easy as it sounds; large or remote sites are excluded because of the time needed, which conflicts with the class schedule and because it is rare and not suggested to work with children on weekends. In addition, a remote or large site may require overnight stays, which increases the cost because of travel and accomodation expenses. Furthermore, the sites must be easily accessible and as safe as possible; in most cases plain valleys are chosen for safety reasons.

It's not rare that despite the fact that many children would like to participate to a field experiment, anxious parents prevent and discourage them. This has to do with the stereotype of an experimental scientist or an exploring geophysicist who is, unfortunately, identified either to an unsociable introvert or to a fearless "Indiana Jones". Thus, some introductory courses and seminars may be attended by worried parents, in order to prove them that field work under certain circumstances may be safer than practicing a sport.

## Kneading a researcher

Working on the field, requires from the young researcher a very well planned preparation and the ability to face and overpass difficulties which may occur at any moment. The students have to prepare a research plan before they get into action. This plan is usually their opinion on a typical set of questions, at which every researcher in geophysics has to answer before performing an experiment:

- Why do you select this specific area?
- Which is the most appropriate method (seismics, electrical soundings, magnetic, ground-penetrating radar, electromagnetics) for your target?
- Which is the most appropriate arrangement in order to detect the target (i.e. spacing between geophones-shots, total length of the whole system)?
- Which is the most appropriate source (hammer, dropping weight), what electrical potential are you going to use?
- Are you sure about the quality of your data? What is the ratio of sound (good data) versus noise (bad data)?
- Is your data saved?
- Is your equipment safe and well packaged?
- Did you take GPS measurements of the shot-receiver points?

Depending on our needs, e.g. what kind of target we would like to detect, at which depth we would like to have an image of the subsurface or what is the desired resolution of this image, the arrangement, the method and the tool differ. It's on students to choose the appropriate work plan, under of course the guidance of a more experienced geophysicist. Back to class, students have to see whether their plan was wrong or right. First, they will have to encounter raw data, i.e. in the form of seismograms as shown in (Fig. 1.).

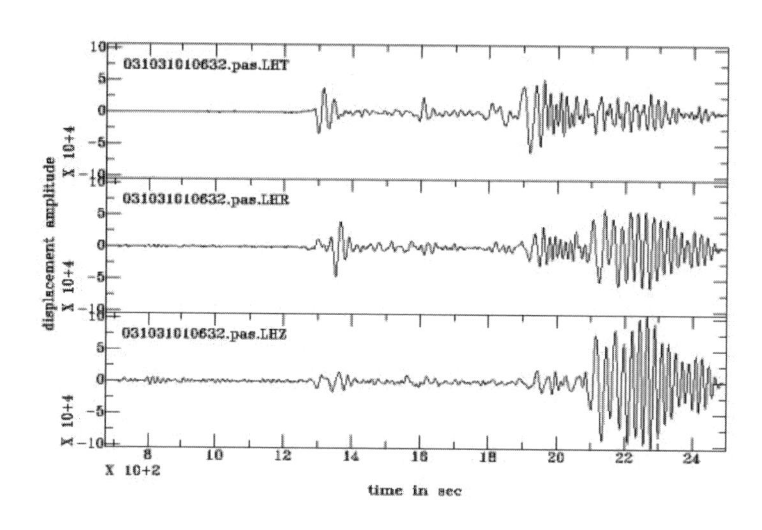


**Figure 1.** A seismogram, showing the time of an event (earthquake) and its amplitude. From this students have to "pick" the first-arrivals, or first-breaks, in other words the starting time of an event. The "picked" times are going to be used in an inversion algorithm in order to provide us with an image of the subsurface. This method is called in general seismic (as the source are earthquakes or shots) traveltime (as the important data is time) tomography.

After picking times, students have to select an initial model which has to be close to the real conditions. This is a great challenge for students, as they have to practice their ability to estimate and to employ in an effective way their theoretical background. This initial model is crucial as it determines the final model, which is the tomogram, the picture of the under-investigation area (Fig. 2.).

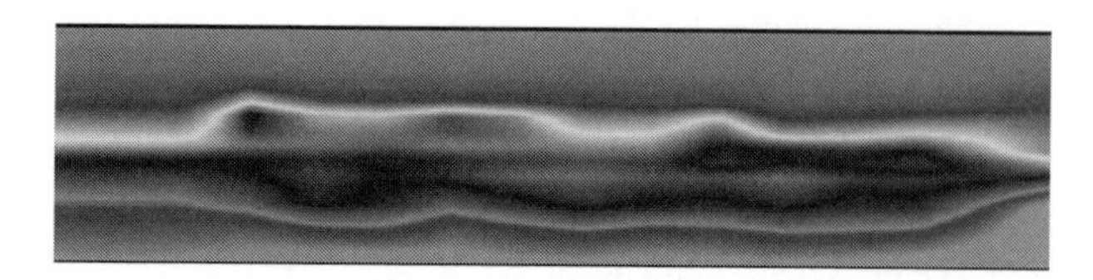

**Figure 2.** A tomogram of the final model, a picture in other words of the subsurface.

## Conclusions

Field-work can help students come closer to the mentality of a researcher. They learn how to prepare a research plan and how to overpass difficulties and obstacles in order to obtain the best possible results. Theory is the tool to manage the plan and it is on them to choose how they will use it.

**References:**

[1] S. Tisseron, Comment l'esprit vient aux objets. Aubier, Paris, 1999.
[2] A. Witten, An eventful semester teaching field geophysics, *The Leading Edge* **7,** (2003) 616-620.
[3] M. Arvanitis and V. Karastathis, Uncertainty analysis of current 3D refraction tomography algorithms on shallow depth applications, *Geophysical Research Abstracts* **5,** (2003), 5284.

*P. Csermely et al. (Eds.)*
*IOS Press, 2005*


# Euroscience's Activities in Greece

**Michael S. ARVANITIS**
*Euroscience Greek Regional Section,*
*PO Box 3125, 10210, Athens, Greece*
*mike@georasis.gr*

**Abstract**. This paper presents briefly the origins and the first steps of Euroscience, the way it is managed, based on the system of regional sections, and the activities of the recently established Greek regional section.

## Introduction

Euroscience is a pan-European, trans-disciplinary organization founded by Claude Kordon and Françoise Praderie back in 1996. The need of a rolling up of the scientists in order to provide a European perspective for science policy, of a union which would represent them, was the fundamental idea behind Euroscience. Such an association did not exist and European scientists used to work in a very fragmented way. The basic memorandum of Euroscience was published in *Nature* in November 1996, indicating the scope and the aims of the organization; to this, positive responses were received from twenty-five European countries, inaugurating a pan-European scientific organization, including many eastern European countries.

The first governing board which was elected in 1998, representing thirteen different countries, decided Euroscience to become an independent voice of science in Europe and to promote scientists' societal and cultural responsibilities. At the beginning thirteen regional sections were launched, the first one based around CERN in Geneva. Among its' activities, workgroups were set up on various topics; today eight working groups exist. The workgroups are the following:

- Science policy
- Integration and collaboration in European Science
- Science and urgent problems of society
- Ethics in Science
- Public awareness of science
- Young scientists
- Technology transfer
- Research and Development in Industry

In addition, conferences and "science cafés" were scheduled but also symposia, based on invited keynote lectures, were prepared for each General Assembly, like "Science and Technology in the new Europe" in Strasbourg (1998) and "Limits and limitations to research" in Freiburg (2000). More recently the Euroscience Open Forum (ESOF 2004) which took place in Stockholm in August 2004, provided a showcase for the association's varied activities.

**Euroscience in Greece**

The regional section of Euroscience in Greece was established just recently, in the summer of 2004. Despite the significant development of the economy during the last years, science and research in Greece are equal to a burden, for both the public and the private sectors, rather than a propitious perspective for the future. Scientists in Greece encounter very often bureaucratic obstacles (i.e. payment delays, procedure vagueness) but also have to continue their research with very limited funds; in 1999, only 0.67% of the GERD was linked with research and scientists support. In addition, private sector seems indifferent to research and prefers to use an innovation rather than innovating. As a result Greece has an increased level of brain-drain, especially to the US and more and younger people are discouraged to continue research after their basic studies.

The Greek regional section had to fight at the very beginning with these obstacles and of course had to bear in mind the important geopolitical role of Greece in S.E. Europe, the Greek scientific tradition and the new emerging industrial fields, like software, plastics, agrifood and maritime industry. As a non-profit, non-governmental organization, the Greek regional section faced people's precariousness in what it was able to do. Fortunately, their fears have not been verified.

In a very short period of time, the Greek regional section managed to enhance links with major research institutes and scientific foundations in Greece and Europe. At the moment, the Greek regional section has close collaboration with universities, students' unions, the government and private foundations. During the first general assembly the scope of the organization has been determined; summing up, its aim is the promotion of science to all levels of society, to facilitate the diffusion of technology, to strengthen the links between Balkan institutes and foundations and the regular publication of an interdisciplinary scientific magazine in Greek.

Special care is given to people with disabilities who are in research or would like to get involved with it. A workgroup was formed, named MFD, which stands for Mobility For Disability and deals with problems, obstacles and best practices for researchers, scientists and students with disabilities. At the moment the workgroup has members from universities and research centres as well as from non-governmental organizations, scientific foundations, disabled unions, professionals from the health care scene, psychologists and doctors.

Another priority of the Greek regional section is the demystification of science in the eyes of children. For the coming winter, many events are scheduled in this direction. Some of these are as follows:

- The concept of "Science - Theatre". Short plays, specially written for small children, in order to narrate the most important moments of science, will be performed on stages all over Greece. Both texts and scripts will be easily accessible by the audience and at the end of the performance a short discussion may follow concerning the topic of the play.
- The concept "Teach your dad, teach your mum", addressed to kids between five and eight, is a series of specially designed books which combine plot, game and learning. The aim is that children will try to "teach" their parents, in a form of game, and, in order this to be realised, they have to "expertise" on a certain scientific subject.

## Conclusions

From its very beginning, Euroscience aimed at promoting science and becoming the voice of scientists in Europe; this, at least partially, has been accomplished. Euroscience now is a leading scientific organization, covering geographically the whole Europe, being responsible for many important scientific activities (i.e. Open Forum, Rammal award, Science Writers Award). The Greek regional section, although recently set up, aims to establish a strong pole in the scientific scene of Greece, to promote science especially to children and people with disabilities, to make science accessible by all, to promote technology and innovation and to ameliorate research conditions in Greece.

**References:**

[1] C. Kordon, The origins of Euroscience: a brief personal recollection. Euroscience News 25 (2003), 2-3.

[2] M. Arvanitis, Mobility in Greece: feasible or problematic?, Early Stage Researchers Mobility 2004, Lisbon, 2004.

[3] S. Anguelov, Euroscience encourages regional scientific co-operation, Euroscience News 26 (2004), 3.

*Science Education: Best Practices of Research Training for Students under 21*
*P. Csermely et al. (Eds.)*
*IOS Press, 2005*



# The Irish Centre for Talented Youth: Motivating Young People towards Science

**Sheila GILHEANY**
*The Irish Centre for Talented Youth,*
*Dublin City University, Dublin 9, Ireland*
*sheila.gilheany@dcu.ie*

**Abstract**. Conducting potency tests on penicillin, discussing rocket technology with a NASA astronaut, analysing animal bone fragments from medieval times - these are just some of the activities which occupy the time of students at The Irish Centre for Talented Youth (CTYI). The Centre of identifies young students with exceptional academic ability and then provides services for them, their parents and teachers. This paper highlights the work of the Centre particularly in relation to nurturing and developing interest in the sciences at an early age.

## 1. Background

CTYI was established in 1992 at Dublin City University in close co-operation with the Center for Talented Youth at the Johns Hopkins University, Baltimore, USA. Its main activities include:

- Talent Identification for students aged 6-16 years
- Specialist Saturday classes from Oct-April in Dublin and 6 other regional centres
- Summer residential programmes in Dublin and non-residential regional summer classes
- All year round correspondence courses
- Teacher Training
- Support for parents
- Research

In addition the Centre also operates and manages the Pfizer Science Bus. This is a mobile lab which travels around primary schools bringing interactive science to over 11,000 children annually. This particular project, while not specifically directed towards exceptionally able children, uses the same teaching approaches as with the rest of the CTYI projects of having every student work to their individual highest possible level.

Estimates for the number of exceptionally able children in Ireland vary depending on the exact definition used. However, taking the Irish Dept of Education's figure that children at the $97^{th}$ percentile level are considered to be gifted, then this suggests a figure of around 23,000 children within the current school population. CTYI works with around 3000 children each year.

It is not uncommon for such students to go unrecognized in school. Some students deliberately hide their ability in an effort to 'fit in', while others learn in unconventional ways which are not always appreciated by school teachers. There is also a sizable group who are very bored at school and literally switch off during school hours. Such frustration can lead to considerable emotional and behavioural problems with students becoming exceptionally unmotivated. CTYI's system of Talent Identification frequently highlights children who had not previously been noted for exceptional ability.

CTYI aims to provide a highly stimulating academic experience in an atmosphere supportive of both social and emotional needs. For many students, CTYI provides the first opportunity for students to meet others who are like themselves and who share common interests. Such support can help give students confidence in themselves which is of enormous value on returning to their usual environment. It also aims to introduce students to academic areas which would not normally be available to them at a young age. This is particularly valuable in relation to the sciences, given the declining numbers of students taking physical sciences at university level.

## 2. Assessments

Students establish their eligibility for the CTYI programme by first taking assessment tests in the areas of mathematical, verbal and abstract reasoning with the Centre and achieving certain age dependent scores. Each year the Centre carries out an annual Talent Search which is a systemic trawl for high ability students in the country with details being sent to all primary and secondary schools in Ireland, all teachers' centres and libraries etc. Numbers taking assessments at the Centre are illustrated in (Fig. 1). Since CTYI's inception in 1993 over **14,600 students** have taken part in an assessment at the Centre with approximately **80%** of all secondary schools having at least one student on the programme. It is interesting to note that for many of the most exceptional students on the programme, their initial identification through the Talent Search was the first formal recognition of their ability. This includes students aged 14 and over who had already been in education for 9 years.

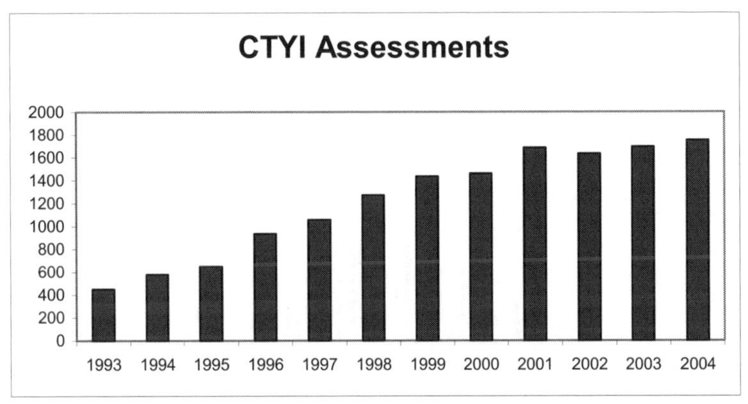


**Figure 1.** Number of students taking assessments at CTYI between 1993-2004.

As a guide to the standard of the program, 13-year old students work at first year university level. All courses are designed to allow the students to work at the pace and depth most appropriate to their ability.

## 3. CTYI Eligibility Criteria

CTYI uses a system of 'out-of-level' testing. In this students take a test which would more usually be taken by students older than them – eg 13 year olds take the USA college entry test which is taken by 17-18 year olds. (Scholastic Aptitude Test). To qualify for the programme they are required to achieve the average score of the older age group. In relation to their own age group this qualifying standard is as follows:

| | |
|---|---|
| ***6-12 year olds*** | *Demonstrate ability at or above the 95th percentile level* |
| ***13-16 year olds*** | *Demonstrate ability at or above the 99th percentile level* |

The assessment results annually show a very consistent pattern, with the average score of participants aged 13-14 years being comparable to the average score of 17-18 year olds on the same test. This is the minimum standard required for qualification for CTYI course. However the assessment results annually also show a small number of students reaching exceptionally high scores. eg 13 year old students scoring over 700 points in the Scholastic Aptitude Test (SAT I) in both the maths and verbal sections, see (Fig 2.). Such scores are only achieved by approximately 1% of college-bound seniors thus demonstrating the usefulness of the SAT as a means of identifying talent of a quite extra-ordinary nature in 13 year old Irish students.

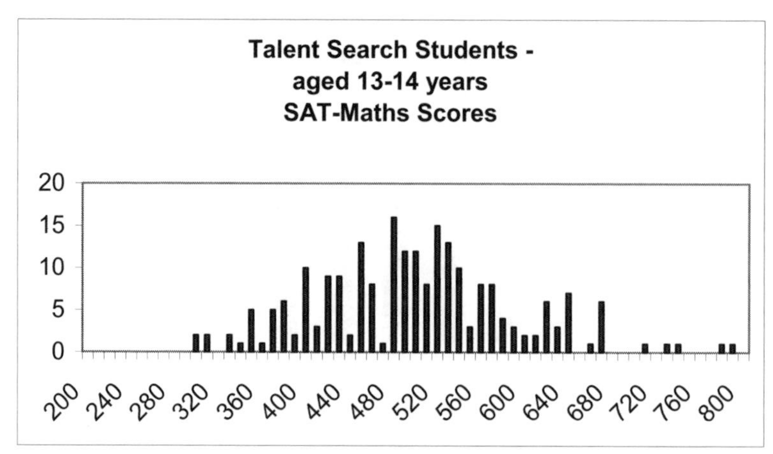


**Figure 2.** SAT scores achieved by 13-14 year old students in 2004.

At primary level around 70% of students taking assessments will qualify for academic programmes while at secondary level approximately 40% of students will reach the required standard. These figures reflect the large degree of self-selection for such programmes. All secondary students who participate in assessments regardless of the outcome are invited to a number of events such as Discovery Days, as an acknowledgement of their ability and desire to seek additional challenge.

## 4. Academic Programme Numbers

At present over 3000 students annually take part in CTYI courses as illustrated in (Fig. 3). Saturday classes are given in Dublin, Cork, Limerick, Letterkenny, Athlone, Waterford and Galway with around 700 children each week taking a course. There is considerable demand from parents to extend the regional provision of courses to other areas. The need for such services is very apparent when it is noted that children are travelling from locations over 100km from the regional centres on a weekly basis. Residential summer courses for 12-16 year olds are given in Dublin while correspondence courses are also available in a range of subject areas throughout the year. It is interesting to note also that students tend to return to the programme repeatedly. Generally in any given course, 80% of the students have previously taken a class with the Centre and return either to take a similar course being offered at a higher level or to try something completely new.

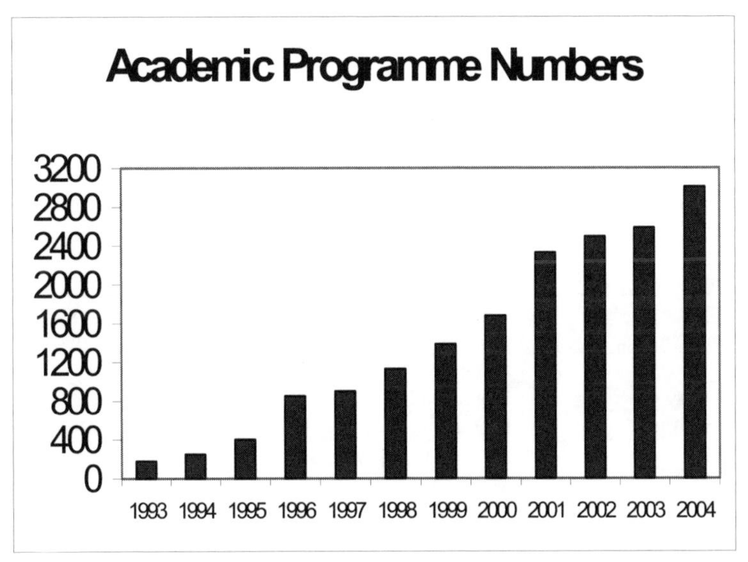


**Figure 3.** Numbers of students taking part in academic courses at CTYI between 1993-2004.

## 5. Academic Courses

Residential and commuter summer courses, Saturday classes, intensive study weekends, Discovery Days and correspondence courses all form part of CTYI's work with young people aged 6-16 years. Saturday classes run over 9-week terms with students studying one subject per term for 2.5 hours per week. Courses are designed and delivered by experts in the particular fields with subjects ranging from Archaeology to Zoology available to students as listed in (Table 1). Such courses normally cover material outside the usual school curriculum. Typically 13 year olds work at first year university level. However, it is not unusual to have students work at even higher levels, including postgraduate type research.

Summer programmes are perhaps the highlight of the Centre's activities. During June and July students aged 12-16 years stay residentially for a three-week period, taking classes for 7 hours per day which run alongside a highly structured social programme designed to support the students' social and emotional needs. Such programmes have a very high student:staff ratio of 6:1. The intense nature of the programme means that students are required to put in a very high level of work during the period and for many students this is perhaps the first time that they have ever really been challenged academically. This can be quite a shock to them but it is usually the case that when they get over the initial surprise, they very much enjoy the satisfaction of working at their full capacity.

**Table 1.** Courses.

| | | |
|---|---|---|
| Advertising | Computational Linguistics | Media Studies |
| Aeronautical Engineering | Computer Applications | Medical Technology |
| Anatomy | Corporate Business | Meteorology |
| Ancient Civilisations | Creative Writing | Neuroscience |
| Archaeology | Electronics | Number Theory |
| Architecture | Engineering | Pharmacology |
| Art | Forensic Science | Philosophy |
| Art Design | Genetics | Probability |
| Art History | Global Economics | Psychology |
| Astronomy | International Relations | Russian Lang.&Culture |
| Atomic and Nuclear Physics | Japanese Lang.&Culture | Science Communication |
| Biology | Journalism | Screen Writing |
| Biotechnology | Legal Studies | Sports Law |
| Business Studies | Linguistics | Sports Psychology |
| Celtic Studies | Literature and Drama | Theoretical Physics |
| Chaos Theory | Marine Biology | World Geopolitics |
| Chemistry | Marketing | Zoology |
| Chinese Lang. &Culture | Mathematical Magic | |
| Classics | | |

It is interesting to note also, that by far the most popular courses on offer to students are lab based science courses. This is particularly the case with students at primary level (6-12 years) with courses tending to be heavily oversubscribed and is in sharp contrast with the declining interest in students taking science at school and university in Ireland. Experience with the students suggests that the highly practical nature of the material being offered is most attractive to them plus the opportunity to work in areas with cutting edge research.

## 6. Schools Outreach

CTYI has a large outreach programme with schools which covers both teacher training in the area of highlighting the needs of high ability children and also school curriculum support. One of the largest of these projects is the Pfizer Science Bus. This is a fully equipped mobile lab, which travels to primary schools around Ireland. Children come on board the Bus and carry out a range of experiments designed to enhance their interest and enthusiasm for science. The project also acts as a support to teachers in relation to the recently introduced science at primary level curriculum. Since its launch in March 2000, this project has met with enormous interest and enthusiasm from both schools and the wider public with over *44,000* children having been on board. Although not specifically targeted at exceptionally able children, it uses the same teaching principles of CTYI that encourage all children to work at the level that is appropriate for them.

Since the launch of the Bus, there has been overwhelming interest in the project, with schools typically seeking to rebook the Bus immediately. All schools are asked to complete an evaluation form. The results from this evaluation process strongly indicate an exceptionally positive reaction to the Bus, with 100% of respondents being highly satisfied with a visit and seeking to have the Bus return to the school. The most typical comments from children are:

***'I want to be a scientist when I grow up!'.***

## 7. Research and Evaluation

Detailed evaluation studies have formed a part of CTYI's work since its first programme. At the Talent Search stage all participants are invited to complete a detailed questionnaire which examines areas such as their attitude towards their ability, school, particular subject areas, career ambitions and perceptions of others towards them. After every course, detailed evaluation studies are carried out with all students, the information from which is used when planning further courses. Lastly a medium term follow up study is being conducted with students from the first three years of the programme. This examines college choices, career paths and perceptions of the impact of the programme on them to date.

A brief summary of some of the data collected to date suggests that:

- 82% of students felt more comfortable with their ability after CTYI course
- 95% - very satisfied with academic challenge of CTYI
- 99% - wished to return for further courses
- CTYI experience influenced 80% in choice of college course

This last factor was frequently sited by students when taking a course with CTYI. In other words, although the students could not obtain course credit for the CTYI programmes, they were keen to study a wide range of courses in order to help them to decide what path to follow at college. Academically they were challenged by the courses and as a result felt more inclined to put in greater effort with their regular school-work as well as wanting to seek out additional challenge. This was particularly the case in relation to science subjects with students citing the benefits of having being exposed to research standard labs and introduced to the techniques of high level science. Follow up research with students several years after their participation indicated these positive benefits were still in place.

P. Csermely et al. (Eds.)
IOS Press, 2005


# The Hands-on Science Project

**Manuel F. M. COSTA**
*Universidade do Minho, Departamento de Física,*
*4710-057 Braga, Portugal*
*mfcosta@fisica.uminho.pt*

**Abstract**. Aimming the promotion of hands- on experimental teaching of Science as a way of improving in-school scientific education and Science literacy in our society, the "Hands-on Science" network was established, in the frames of the action Comenius 3 of EC' program Socrates, in October 2003, by twenty-eight institutions from ten European countries (BE, CY, DE, ES, GR, MT, PT, RO, SL, UK) and a transnational consortium (CoLoS). The H-Sci network has a broad remit, aiming to promote and diffuse among schoolteachers, schools, and national and transnational educational boards, well established and newly investigated practices of hands-on experimental teaching of Science in all its disciplines. We will do this by fostering the development and use of hands-on experiments in the classroom so that students "do" science rather than merely being "exposed" to it.

## Introduction

In the Society of our days there is a major increasing need of an in depth quality education in Science and Technology. Science teaching at school should be generalised aiming not only the sound establishment of a "Science" culture in our societies but also to guarantee a steady basis for the improvement of Science and its technological applications.

Urgent actions should be taken in this direction. By initiative of the author the "Hands-on Science" (H-Sci) network was created. The European Commission under the program Socrates, Comenus 3 action (project nº. 110157-CP-1-2003-1-PT-COMENIUS-C3) supports the network.

The activities of our network focus on the diffusion and development at European scale of positive hands-on experimental practices on teaching science at basic secondary and vocational training schools. By leading the students to an active volunteer and committed participation in the teaching/learning process through hands-on practice and experimentation, making intensive use of the new instruments and resources of the Information Society.

Hands-on practices are being very successfully employed in Science Museums' Clubs and Associations throughout Europe, as major motivational but also learning tools. The benefits of the use of these methods in the classroom are being illustrated for already some years in the USA and in some EU schools. It is now the right time to generalise the use of these practices in an effort to raise the levels of scientific literacy and expertise among our youngsters.

Science and Technology have always been issues of fundamental importance to the development of the countries and societies. Today this is, definitively, also true. No sound development can be foreseen without a strong S&T infrastructure. The fundamental basis of it is, indubitably, a well-prepared motivated reliable and flexible workforce. Scientific and Technological literacy is also of great importance for citizenship and democratic participation in a world where Science and Technology issues and demands have a dominant role.

The increase of literacy in Science and related technical expertise will induce the creation or development of high-tech industries or services' firms that are of fundamental importance for Europe's growth and development. Thus being a pro-active factor of direct impact on the inclusion of new, and less developed, country members in the EU. Not only a long-term effect is foreseeable but also an immediate impact by raising the expectations of the economic agents and of the society in general.

A clear picture on the status of Science and Technology Education in European Schools should be made available. As well a number of policy recommendations and tested pedagogic material will be made available. A network of experts on this topic will be established. A raised awareness of the importance of this theme will be established in steady basis.

**Figure 1.** The symbol of the network.

## 1. Pedagogical approaches

The new stringent requirements of the modern society demand not only the gathering of specific knowledge but also and specially the ability of acting interactively to be able to find, analyse and solve new interdisciplinary problems.

The best way of achieving an adequate formation of our students on these issues is by inducing the students to an active committed participation in the teaching/learning process, through practice and experimentation [1]. Making use of the new instruments and resources of the Information Society.

Our pedagogic approach is focused on inducing an effective learning of science subjects and basic competencies as responsibility, method, self control and reward, critical reasoning and observation, commitment in collective group actions, interaction and proactive inter-dependence.

Hands-on activities are proved to be the most effective way of acquiring these competencies for the vast majority of child and adolescents [2]. Whenever advisable a constructivistic [3] approach will be used. It is sometimes necessary to allow the students to have a first qualitative conceptual contact with the issues of science. Conceptual learning will be integrated and used in our hands-on practices. We will use virtual simulations of natural phenomena always simple and inducing or requiring interaction with the students in a way that the qualitative perception of the concepts of Science may lead to a quantitative interpretation. The Physlets [4] first developed at the Davidson College in the USA are accepted to be a good tool that will be explored in this context, and exported to other fields of Science. Different approaches to this issue of the use of ICT including virtual simulations in Science teaching will be discussed and assessed. The use of ICT in teacher training will also be explored [5]. The produced guides will be formative summative and making constant appeal to critical reasoning, observation and active commitment of the pupils. They will be expert' reviewed and tested in-class, and receive students and school evaluation. The heterogeneity of pupils interests, abilities, experiences and future wishes should be carefully assessed also by contrasting the responses between different countries, languages, minority groups, cultures and gender. This analysis, that will be published and widely publicised, with conform the development of the pedagogical materials and syllabus to be produced for different countries, populations and languages. We will try to establish bridges between the industry and employers, the schools and educators, and research institutions. Suggestions will be developed on how to establish this kind of links (local and international co-operation settings will both be considered). Syllabus and guides to be developed will also take into account this interaction.

The assessment of the network is considered fundamental. We will seek both internal self-evaluation but also from external educational government boards, Educators Associations, relevant local government entities, teachers and especially from students and their parents. Inquests and questionnaires will be prepared and delivered to the different evaluators. The experience of the ROSE' (The Relevance Of Science Education) project coordinated by the University of Oslo, will be taken into account [6].

Our pedagogical and organisational strategies follows the guideline determinations of the USA' National Science Education Standards (National Research Council, 1986) actualised and adapted to the actual, and local, social cultural and educational situation in the EU. We place a special focus on the pro-active commitment of teachers and educators in motivating, inducing and conforming the autonomous realisation of broadband hands-on scientific activities by keen and active pupils.

## 3. Main goals and activities

The main expected outcomes of the network are sumarized below:

- Teaching recommendations
- Syllabus proposals
- Web-based discussion forums
- Web-pages and sites
- Virtual laboratories and experiments
- Translated reports and written material
- Guides and pedagogical guidelines in official national languages
- Catalogue of projects
- Catalogue of books and guides
- Info folder about the network
- Electronic newsletter
- Teachings packages
- In-school motivational activities
- Three International Conferences and thematic workshops
- Several training courses Comenius 2.1 project's proposals
- Several Comenius 1 project's proposals
- Internet based video-conferences and meetings in local national languages
- Network of Discovering Science Student' Clubs.
- Network of pattern makers in hands-on learning of Science.

A task force is being established under the umbrella of the H-Sci network to co-ordinate a public relations campaign about Science education. This will be aimed at students, teachers and educators, schools and the community, Ministries of Education and Education bodies, in a systematic attempt to demonstrate the benefits of an extended use of hands-on learning of Science. Interactive web sites and virtual simulation tools and labs with open access to all will be established. Educational kits of hands-on experiments and support material at different levels of complexity will be designed created and discussed. Textbooks and reports, including electronic interactive versions, will be produced in different languages and disseminated.

Changes to the National syllabus in Science that, we believe, will enhance the effectiveness of science teaching, will be discussed and proposed to the proper authorities. Annual international conferences and workshops will be held, in addition to a significant number of training courses for schoolteachers. Meetings and transnational exchange visits will be organised.

New members from all EU countries will be associated to the network. The network will promote and induce new co-operation projects at EU level in the field of Science and Technology education. The establishment of student' Science Clubs will be induced. Science fairs, contests and activities that bring together students, teachers and education specialists, research institutions, the industry and the community in general, will be a notable feature of the network.

We are also planning activities that we expect will have a positive impact in social inclusion of minorities. In particular in what concerns the gypsy community in Romania and immigrants and religious minorities in countries like Germany, Malta or Portugal. In some of our Schools there are a large increasing number of minorities that need special attention. The use of the kind of activities we propose that induces an active voluntary commitment of the students in concrete tasks can have rather positive effect contributing to a better integration of minorities in the School and in the community. Furthermore these activities lead to an organisation of the students in groups were each student will have a common appealing goal.

The gypsy community in Romania will have a particular attention. Lectures and demonstrations will be delivered to those communities in deprived areas in the interior of Romania.

We consider that hands-on teaching of girls in School age may contribute to define a clearer perception of Science on this group of students approaching helping bring more women to Science and Technology. The ways of achieving this goal will be discussed explored and assessed.

Close attention will be given to the possibilities and benefits of hands-on science activities in special education schools for children with disabilities. Pedagogical materials and strategies will be developed and in class tested. We expect to prepare other projects in a near future in order to deal with this specific problem.

The issue of Science literacy connected with the each day more important adults' and life-long education will also be addressed also in the sense that a positive appealing initial or basic contact with science may certainly lead to an effective sustainability of the interest about Science and Technology during life.

We also have as a major general goal to contribute to a faster positive inclusion of the future new EU members in an enlarged Europe.

## 4. Target groups and expected impact

To our knowledge it is the first time such a wide network is established in Europe to address the issue of hands-on learning of Science and Technology. Taking profit of the new ICT technological developments in modern society we will support complement and integrate the Hands-on quantitative study of Science with Constructivism and Conceptual and Interactive Engagement qualitative Learning.

Apart from the members and associated members of the network we are targeting Schools, School' students, teachers and educators, teacher training Institutions, education boards and responsibles of national and trans-national educational programs, and research institutes, the local communities and the industry.

Directely an indirectly the first beneficiaries of our activities will be the Science' schools students not only in the involved countries but also in all EU current and up-coming member countries.

The impact of our activities depend on the target groups:

- The network participants will get a clear idea of the state of the art on hands-on learning of the different subjects of Science.
- The network participants will be able to serve as pattern makers in their schools, communities and countries in the subjects of hand-on experimental learning of Science.
- The network participants will establish an enlarged number of international contacts that will enlarge their perspectives as educators but also as citizens of an enlarged European Union.
- Raised awareness of the importance of international partnership among teachers, schools and governments.
- Schools and educators will have a proven idea of the true advantages and possibilities of the pedagogical use of hands-on experimentation in the teaching and learning of Sciences.
- Pupils will have raised knowledge on a number of topics and will have a new international perspective about science, teaching and education.
- Students of the participating schools (the networks ones and other to be associated) will acquire new competencies and be aware of its importance.

We expect this network to serve as starting point for the generalisation of the use of hands-on active learning of Science in EU Schools.

## 5. Partnership composition

The "Hands-on Science" network involves several Universities, Research Institutions, National Government Boards, Private companies, Colleges and Schools and School' Associations from Portugal, Greece, Spain, UK, Slovenia, Romania, Germany, Belgium, Malta and Cyprus, and the CoLoS (COnceptual Learning Of Science) pedagogic association.

The network counts with the formal involvement of 29 Institutions of different types covering all education levels from elementary secondary, special education, technical and university levels. Also a large number of associate member are already involved from the participating countries and from France, Denmark, Norway, Holland, Slovakia, Austria, Belarus, Bulgaria and Russia (observer). Further associate member will be accepted along the development of the network activities, in specific task forces or thematic groups.

Different institutions working in different environments with distinct cultural and socio-economical backgrounds – in large towns, medium sized towns, and small villages, some in rural or industrial areas- covering a large varied geographical area from the Iberian Peninsula, the Mediterranean, Southern Northern and Central Europe to the Eastern Europe. They will focus their activities in exploring and promoting the use of hands-on experimentation teaching by extensively employing advanced ICT tools, or simple hand made materials, in integrated interdisciplinary approaches or analysing the issues in a more sectioned way. Different approaches, different feelings, the same conscience, the same objective: to evolve to a positive approach of participated active hands-on experimental learning of Science at school level.

All partners have previous experience on exploring and or developing innovative educational practices and materials in a non university level and many are or were involved in national projects of these kinds and on international collaborations within the EU or with EU and North and Latin American' or North-African Institutions.

Some of the participating schools have relatively large (up to 30%) groups of pupils from minority groups (language, race, religion) and are experienced in developing pedagogical approaches to deal successfully with the problems that situation poses. A number of schools are experienced in Distance Learning and most have already proper facilities (some financed through EU projects). Several network member institutions had or currently have members of their staff participating in curriculum development efforts in science and technology organised by the respective Ministries of Education, Science and or Culture.

The CoLoS consortium brings together research teams from many US and European Universities (from the DE, UK, IT, RU, SI, FR, ES, PT). It aims to promote the development of innovative teaching methods in science and technology focusing on: learning and understanding fundamental concepts in science; the integration of qualitative and intuitive understanding with quantitative methods; and the use of simulation and network-based material.

A number of different Schools and Institutions will be involved in the Network' activities as associated members. Among those a few will a role focused on evaluation and or dissemination of the network activities and results. The associated schools and institutions will not only be informed of the network activities, have privileged access to web based network and to the produced reports and materials, as they will be asked to participate in the activities of the thematic groups as extensively as possible.

A numbers of Observer members from countries outside the EU will be accepted and welcomed.

## 6. Implementation strategy and planning of activities

Our network is relatively large either in terms of the number of partners and countries involved. We involve all fields of science all levels of School teaching with an extensive range of general and specific goals.

In order to maintain the network with a manageable but sufficient size in order to fulfil all requirements and develop its activities efficiently, we proceed with a selection of partners on view of their expertise and previous experience in co-operation projects particularly with other members of the network' team. A relatively large group of associated members was established enhancing the impact of our activities.

We decided that the network needed a 3 full year's period of activity to achieve the direct goals and outcomes and specially to guarantee its sustainability in the years after the formal end of the project.

An efficient decentralised and hierarchically organised system of administration of the Network on all its aspects will be established. The network Cooordinator (the author) will ensure the proper efficient development of the network. A Steering Committee formed by the all the National Coordinators and the Network Coordinator will meet (frequently by videoconference) as frequently as needed every year in order to: analyse the development of the work; assessment of results; writing down reports and further support material; proceed with data diffusion and public relations.

In regular basis each National Coordinator will contact all the responsibles of the member institutions of their country recalling their work results. A short report will be issued and delivered to the Network Coordinator that will disseminate it by the entire network members for discussion and improvement. Whenever final conclusions, among the Network, are achieved they will be published in the web site in order to extend the discussion to the community. Boards or offices of the National and supra-national Educational institutions will receive that information directly. Most of the meetings will be (whenever possible) made in the form of web based video-conferencing. Most of them will be open to all Network Members that often will be specifically invited to take part on the meeting. The in person contact between the Network members is considered very important and will be made whenever possible also during the exchange visits and workshops to be organised throughout the different Partner Countries but also on the General Network Meetings that will take place during the Annual International Conferences.

At least once a year of activities the Network Coordinator will travel to Brussels in order to meet the representatives of the Socrates Program to present and discuss results, reports and activities schedule for the following year. The eventual participation of National Coordinators could be considered.

Annual International Conferences will be held on the middle of each project's year and will be attended hopefully by the entire network members. Furthermore it will be open to all scientific and educational community (including students and their parents). The community will be invited to participate not only in conferences and thematic workshops but also in courses and motivating/promoting activities in the classroom or at School.

This network is organised in a top-down/ bottom-up structure. This means that not only the Network Coordinator and National Coordinators will induce actions produce reports guidelines and induce discussion and interaction. Also all the Network Members on his thematic groups or taskforce or individually through his group will present suggestions, initiate discussions prepare reports and organise local or multilateral activities.

The General Network Meeting presided by the Network Coordinator (once a year during the annual conferences) is on the top of the hierarchy of our structures and to which the Steering Committee will report.

The network workforce will be divided into thematic groups covering all fields of Science and fourteen main task forces.

Transversal thematic workshops have been or will be organised: "The access of Women to Science" in Germany in June 2004, "The challenges of EU' enlargement on Science literacy and Development" in Malta, fall 2005; and "Science Literacy and Life-long Learning" in Romania in July 2006.

In January 2004 we launched the 1st European Contest on "Ideas for Science Fairs" with which we hope to induce the generalisation of the organisation of Science fairs in the Schools.

**Figure 2.** The "Ideas for Science Fairs" European contest is expected to involve many hundred students all over Europe.

During each annual meeting a workshop will be organised where we will present our work and discussing it with industries' representative and of the community. Their ideas and demands will be organised and will condition the further development of our work.

Several Comenius 2.1. projects will be presented by network members in order to develop a number of training courses in the different languages and participating countries. Most of them having an informal interactive structure aiming to allow the teachers/educators to feel the advantages of this pedagogical approach. Practical tools will be giving to the teachers in order to allow them to immediately begin introducing hands-on activities in their classrooms. Other topics will surely appear along the way. By the moment and already for the next call we expect to present 5 projects on "Elementary Optics", "The use of virtual interactive simulations in science teaching", "How to establish an In School Science Forum", "Robots. On the way to the future", "Feeling Life and ecology". On these topics and in others like "All different all the same" involving "regular" and special education schools in Portugal, Spain and Greece, or "The world of photonics" different Comenius 1 project proposals will be made.

The schools involved on the project will play the most important role. However in the network close to each one of them we intend to have a higher educational institution working close together. Several motivational activities will be organised in the different countries inside the school in extra-curricular activities like informal seminars, Science fairs and contests, but also as in-classroom intervention were the students will execute a number of hands-on experiments. A general set of experiments covering the different disciplines of Science (ecology, light sound and waves, geology the earth and the environment, biology animals and plants, genetics and reproduction, speed and mechanics, electricity and energy, sun and space, chemistry, shapes and forms, …) and education levels (up to 10 years old, 10-13, 14-16, 16 to 18) will be studied and supplied to the schools for assessment of the materials them selves and different approaches of its practical in class use.

In Romania the local partners with cooperation of members of other countries when possible will prepare and execute a number of informal courses and demonstration sessions in rural and particularly poor and neglected areas, including gypsies communities in the interior of Romania.

Web based meetings and teleconferences will be used as extensively as possible. An ISDN multipoint videoconference system will be established. The use of virtual hands-on learning tools will be explored. Different approaches will be implemented and tested. Translations and adjusts will be made at Country level. The Science Fairs and Students' Science Clubs that will be organised will have virtual versions. A suitable and simply platform will be created allowing a straightforward organisation of virtual web-based Science Fairs and students' a Science discussion e-forum.

The careful assessment of the teachers and students feedback on the Project's activities and product' outcomes will be made and is considered fundamental.

We will seek an enlarged involvement of Industries and of the work-world in general in order to get their feedback on educational and special competencies needs. We also aim to gather a cluster of companies (including major industries that will be invited to attend our meetings and assess our activities and outcomes) able to ensure the development and financing of the networks' activities after the official expected end of the H-Sci network.

## 7. Evaluation and dissemination

The assessment of the network is considered fundamental. Internal self-evaluation will be made in a country and task basis. At least twice a year the Steering Committee will access the evolution of the network activities and establish corrective measures whenever needed. The evaluation will be made in two levels: in what concerns organisational matters, and on the evaluation of the quality of the pedagogical material and activities developed. Reports will be generated and publicised. By the end of each year reports will be published. In the network global web site there will be a space for evaluation and discussion of progress and results open not only to the network members but also to all interested. National web sites, dynamically linked to the global website, will include a space for pupil's opinions and intervention. A data processing strategy will be established during the first months of activities of the network being adjusted as necessary. The strategy will be made available to all interested for use or discussion.

Acquired data will be statistically analysed and conclusions drawn. The network will also request access, and analyse and process the corresponding data, to the current exams and evaluations on the schools and students involved in activities developed by suggestion or authorship of the network. The Project and Projects' activities and outcomes will be widely publicised.

Apart from the International Conferences and workshops the results of our activities will be presented in different conferences and congress in relevant subjects. As well several papers will be published in International Journals.

The opinion of all the students involved on our network activities will be expressed in volunteer anonymous inquests that will be prepared. The results of the inquests will be statistically treat and the main conclusion taking in great account.

## 8. Conclusion

The pedagogical usefulness and effectiveness of in-class hands-on experimental activities is clearly proved in different school levels and disciplines. We intend to further prove and make this evident contributing to invert decline of interest among young people for science studies and careers.

The "Hands-on Science" network aims to strategically induce the creation of a realm of learning that will give the students the competitive edge in the new Knowledge-based Economy. We plan to grow steps in the sense that schools may in fact become incubators for nurturing promising scientists and pro-efficient technicians and professionals in Science and Technology.

## 9. Acknowledgements

The author as co-ordinator of the network would like to acknowledge commitment of all Members of the network. We also acknowledge the support of the European Commission under the Socrates Project nº. 110157-CP-1-2003-1-PT-COMENIUS-C3.

**References:**

[1] Manuel F.M. Costa, "The importance of experimentation on teaching the basics of optics at elementary and high schools", Proc. Soc. Photo-Opt. Instrum. Eng., vol. 3190, p. 228-232 (1997).
[2] "Experimental learning: Experience as source of learning and development";D. A. Kolb, , Prentice Hall, Upper Saddle River, N.J. 1984
[3] "Constructivist Teaching in Primary School Social Studies, Mathematics, Science, ICT and Design and Technology"; Gatt S. & Vella Y. (2003).
[4] "Physlets: Teaching Physics with Interactive Curricular Material", Wolfgang Christian and Mario Belloni, Prentice Hall's Series in Educational Innovation, ISBN 0-13-029341-5 (2001).
[5] P. G. Michaelides, An affordable and efficient in-service training scheme for the Science Teacher, International Conference "Computer Based Learning in Science", Proceedings Volume I pp 792-799.
[6] http://folk.uio.no/sveinsj/ROSE_abstract.htm

P. Csermely et al. (Eds.)
IOS Press, 2005


# Promoting Biochemical Research in the Secondary School

**Josep M. FERNÁNDEZ-NOVELL and Joan J. GUINOVART**
*Department of Biochemistry and Molecular Biology and*
*IRBB, Barcelona Science Park, University of Barcelona, Spain*
*guinovart@pcb.ub.es*

**Abstract.** To bridge the gap between the secondary school and the university, the Department of Biochemistry and Molecular Biology of the University of Barcelona has designed two course programmes, both of which include lectures and practical classes and aim to promote biochemical research. The first addresses final year secondary school students and intends to increase their interest in biochemistry. The second programme has been designed for secondary school science teachers and aims to update their knowledge of biochemistry and provide them with a booklet of laboratory practices. These courses have been successful in their objectives and are now reference models for other disciplines.

## 1. Introduction

Knowledge in biochemistry and molecular biology (BMB) has grown exponentially in recent years. Words like DNA, transgenic, genome, stem cells, prions or AIDS are frequently present in the "news" and the language of young people. However, most secondary school students do not understand these terms and do not appreciate the contribution of biochemistry and molecular biology to our day-to-day lives, for example to health, the environment, the economy, etc. As a result, secondary school teachers are facing an increasing number of questions by their students on the meaning of these words and their context in biochemistry.

In Spain's secondary schools, few concepts in biochemistry and molecular biology are explained and these are only given in biology and chemistry courses. Moreover, the number of laboratory experiments is particularly low. Furthermore, many science teachers in secondary schools do not hold degrees in Biology and/or graduated around 1970-1980 when university curricula did not include Molecular Biology.

In Spain, university science degrees are divided in 3 periods (cycles), biochemistry is a two-year second cycle degree and students must previously complete a first cycle (2-3 years) in chemistry, biology, pharmacy, medicine or veterinary medicine to enroll. Because biochemistry is included in the second cycle, secondary school students are given little or no information regarding this option.

To stimulate secondary students' interest in biochemistry and to increase the knowledge of their teachers in the new developments in this field and to improve the experiments performed at this level of education, the Department of Biochemistry and Molecular Biology of the University of Barcelona [1], in collaboration with The Spanish Society for Biochemistry and Molecular Biology and The Catalan Board of Teachers (Generalitat de Catalunya-Regional Government) [2], has organized two course programmes, one addressed to secondary school students [3] and the other to their teachers [4]. Moreover, a handbook on experiments has been published [5].

## 2. Biochemistry course for Secondary School students

The course "I ♥ biochemistry" started in 1997 for students in their final year of secondary school and has now been running for 8 years. The one-week course aims to introduce these students to biochemistry, to increase their research vocation and to orient those interested in biochemistry and molecular and cellular biology.

### 2.1. Student selection

A total of 4,686 applications have been received over these years. The number of yearly applications has now stabilized at between 400 to 500. The applicants are requested to write a letter stating their interest in the course and this must be accompanied by a recommendation from their science teacher. These two documents together with the students' academic records are used to make a first selection, which reduces the number of candidates to around 100.

On one Saturday, the 100 selected applicants are invited to laboratories at the Dept. of BMB to perform a number of experiments which are supervised by PhD students and Faculty members. In addition, the candidates are interviewed to establish their level of motivation and interest in BBM, after which the most promising 24 students are invited to do the full course.

### 2.2. The course

Each day of the course includes laboratory classes and lectures. Before commencing laboratory practice (DNA extraction, glucose concentration, internet and biochemistry…), an extensive explanation, scientific fundaments, use of instruments and a protocol of each experiment is given by PhD students. The discussion of results and procedures is stimulated at all times.

Lectures (protein DNA and RNA, metabolism, biotechnology, gene therapy, molecular pathology,..., the most important aims in biochemistry and molecular biology) are delivered by professors of the University of Barcelona and other universities.

Over informal lunch breaks, students have the opportunity to ask questions, talk about research in the university, and discuss science related topics with their instructors (normally PhD students), Faculty members and lecturers. Students are also advised on the academic options that lead to a degree in Biochemistry.

Part of the final day is devoted to stimulating the analytical capacity of students through the presentation, comparison and discussion of experimental results.

### 2.3. Course Feed-back

Student feedback is collected through a questionnaire. This survey indicates that students greatly appreciate the opportunity to perform experiments as these are not compulsory activities in secondary school curricula. A very good interaction between the students and the lecturers and the personal interaction with the instructors are also marked.

89 % of students who participated in the 8 courses have been followed-up to check their study choice and i) most chose degrees, first cycle, with a considerably presence of biochemistry in their programmes, ii) 27 % had enrolled on a biochemistry degree and iii) 14 % had enrolled on a doctoral degree in biochemistry (PhD).

## 3. Biochemistry course for Secondary School teachers

The 30-hour course "Updating in biochemistry and molecular biology" was started in 1998 for secondary school science teachers and has been running for 7 years.

Many of these teachers in Spain have a limited knowledge of BMB because some did not graduate in biology or, particularly for senior staff, molecular biology concepts were not available at the time they studied their degrees. Furthermore, most teachers limit their programs to "classic" chemistry and biology and rarely give laboratory classes that address modern biochemistry. Therefore, their students frequently have little awareness of this field of science. Furthermore, secondary school curricula in Catalonia now include a compulsory research project that must be done by each individual student. Given that these projects are supervised by science teachers, they now require up-to-date information about attractive topics that could be addressed.

The course aims to consolidate knowledge in biochemistry, to provide science teachers with a set of biochemical and molecular biology experiments that can be used in their laboratories and to propose ideas on potential research projects for the students.

### 3.1. Teachers-students

In the 7 years that the course has been running, 24 secondary teachers have enrolled every year. 79 % of participants held a degree in Biology; 15 % in Chemistry, 2 % in Pharmacy, 3 % in Geology and 1 % in Veterinary Medicine. Participants were on average 48 years old and they graduated around 1980.

### 3.2. The course

The course includes lectures and practical classes. Lectures, given by Faculty members of the Dept. of BMB at the University of Barcelona and invited speakers from other universities, cover areas such as PCR, gene expression, DNA chips, biotechnology and the most important current applications of biochemistry. Examples and discussion with lecturers increase secondary teachers' interest in biochemistry and provide them with the answers to their students' questions about the current topics in BMB.

Laboratory classes are given by PhD students and Faculty staff. These classes give an extensive explanation, scientific fundaments, use of instruments and a protocol of each experiment.

Participants, in small groups, use restriction enzymes to manipulate DNA, ELISA, PCR, cellular culture and chromatography procedures, among others. The discussion of results, procedures and also conclusions is stimulated at all times.

Participants are provided with a booklet on biochemistry and molecular biology research projects in secondary schools, which is discussed on the last day of the course. Dietary fat, war on bacteria, flies and genes, plant proteases or the chlorophyll around the sun are some of the projects covered in the booklet. They are developed as a scientific paper, and therefore include objectives, introduction, material and methods, results, conclusions and references. At the end of each project, there is a list of potential spin-off studies, thereby simplifying the teachers work by providing ideas on research topics for their students.

### 3.3. Course Feedback

A questionnaire is used to obtain an overall evaluation of the course. The results indicate that secondary school science teachers appreciate the opportunity to bring their knowledge on this scientific field up to date and highlight the good working relationship established between lecturers, PhD students and Faculty staff.

A few years after taking the course, participants are asked to complete another questionnaire which assesses the changes incorporated into their teaching of biochemistry and molecular biology.

The results obtained show that i) 70 % of the interviewees had incorporated the new information into their theory classes and 23 % planned to use this information in the future, ii) 48% of the interviewees had incorporated the new information into their laboratory classes and 34 % planned to use this information in the future, iii) 86 % expressed that the course had completely changed the explanations that they gave in BMB classes, iv) 100 % had used the research project booklet.

## 4. Conclusions

Secondary school students and their science teachers have received the courses with great enthusiasm. These courses promote research in secondary schools and are good examples of the relationship necessary between the University and the secondary school and highlight on the one hand, the educational facet of the former and on the other, the interest in increasing knowledge and research in biochemistry and molecular biology in the latter. Moreover, the courses have now become reference models for other disciplines.

**References:**

[1] M. Martínez, B. Gros, T. Romaña. The problem of training in Higher Education. Higher Education in Europe, vol. XXIII, n. 4 (1998) 483-495.

[2] A. Corda, T. Ruzzon, S. Lercari, S. Ucelli. The role of scientific institutions in the promotion of Biotechnology to the public (school, the mass-media, entrepreneurs etc). Biochemical Education 26 (1998) 52-55.

[3] J.M. Fernàndez-Novell, R.R. Gomis, E. Cid, A. Barberà and J.J. Guinovart. Bridging the gap in biochemistry between secondary school and university. Biochemistry and Molecular Biology Education. 30 (2002) 172-174.

[4] J.M. Fernàndez-Novell, E. Cid, R.R. Gomis, A. Barberà and J.J. Guinovart. A Biochemistry and Molecular Biology Course for Secondary School Teachers. Biochemistry and Molecular Biology Education. (in press)

[5] J.M. Fernàndez-Novell, R. Fusté and J.J. Guinovart. Temes de Bioquímica. Treballs de recerca. Ed. Universitat Barcelona, Barcelona 2000.

P. Csermely et al. (Eds.)
IOS Press, 2005


# Bioinformatics for All: Scientific Outreach at the European Bioinformatics Institute

**Catherine BROOKSBANK**
*EMBL-European Bioinformatics Institute,*
*Hinxton, Cambridge CB10 1SD, UK*
*cath@ebi.ac.uk*

**Abstract**. The advent of genome sequencing heralded a new era for the biological sciences. Previously we spent most of our time in the lab generating data, whereas now we are now awash with data and our efforts must go into finding more efficient ways of analysing it. A new field, termed bioinformatics, evolved to meet this need. The European Bioinformatics Institute is an important hub of research, service provision and training in bioinformatics. As such, we take very seriously the need both to inform the general public and educate the biologists and bioinformaticians of the future about how bioinformatics and the new biology will affect all our lives. Here I discuss some of the outreach activities that we've been involved in in the past two years.

## Introduction

Scientists have invested a huge amount of time and money into deciphering our full genetic sequence and those of many other organisms. Why? DNA holds the key to how every organism on the planet functions. If we know the sequence of all our genes, we can predict the sequence of all our proteins. We can begin to work out how tiny changes in the sequence of a gene can have profound effects on the function of the protein that it encodes. And by understanding the nature of these changes we can attempt to find rational ways of preventing and treating the diseases that these changes cause. Similarly, understanding the genomes of other organisms is finding application in agriculture, food manufacture, and environmental science. It is therefore vitally important that students, irrespective of whether they intend to pursue a career in science, leave school understanding how this new biology will affect all aspects of their lives – from the diagnosis and treatment of disease to their food supply.

*Our responsibility is an educational opportunity*

The European Bioinformatics Institute (EBI, www.ebi.ac.uk), which is a part of the European Molecular Biology Laboratory (EMBL, www.embl.org), fulfils a vital public service by making a treasure-trove of biologically important information publicly accessible. This information comes from genome-sequencing projects and other high-throughput projects for gathering molecular biological information (for example protein structures, protein–protein interactions). We do this because we firmly believe that freely available information will accelerate the advancement of science.

The main users of our data resources are, of course, biological scientists. Bioinformatics is a word that few school students or members of the general public have encountered; should we be spending time trying to preach such a specialized subject to them? I would argue that the EBI's policy of making biological information freely available also provides a wonderful educational opportunity.

In the past two years we have been involved in a number of projects that have brought the the fields of genomics and bioinformatics to pupils, teachers and the general public. The reaction from our participants has been extremely positive: they are often surprised and delighted that they can explore the same data that scientists use on a daily basis to advance our understanding of biology. Here I outline three different events that the EBI has been involved in, and try to draw some general conclusions about how to organize successful outreach events.

## 1. Exploring your genome: a treasurehunt on the Internet

In summer 2003, EMBL celebrated 25 years in Heidelberg – the home of its central laboratory. As part of the celebrations, it threw open its doors to the populace of Heidelberg. Activities included hands-on practical classes, tours and a science café in which participants could browse exhibits about the laboratory. The EBI organized 'Exploring your genome: a treasure hunt on the Internet', as a way of introducing people to the human genome, why it's been sequenced, what we can learn from the sequence and how genomic information can be linked to other types of biological information to help us learn more about ourselves – in health and disease. The entire activity lasted about 80 minutes; it was held in German and repeated four times throughout the day.

The activity began with a 20-minute introductory presentation that explained what bioinformatics is, the basic biology needed to do the treasure hunt (a primer on the central dogma – DNA makes RNA makes protein), and an explanation of how the treasure hunt would work.

The treasure hunt itself focused on finding out about a mystery gene. Participants (grouped in pairs with one computer per pair) were given a short DNA sequence and asked to find the three bases that code for the start of a gene (the start codon). They then had to translate a short stretch of the genetic code into the protein code (they were given a 'codon wheel' to allow them to do this); next they were shown how to do a homology search to find the protein whose sequence they'd translated. They then used several EBI-based databases to find out which chromosome the mystery gene was on, what disease results if there are mutations in the gene, and some structural features of its protein product. Each participant was given a printed workbook in which to write the answers. All the information needed to do the practical was printed in the workbook, and the only other equipment needed was a computer with an Internet connection. Prizes were awarded to all participants, but the team that finished first got an extra prize.

The aim of our activity was more fundamental than to give the participants a taste of bioinformatics and its uses. We tried to use bioinformatics as a vehicle to illustrate some important biological principles, such as the elegant simplicity of the genetic code, the link between genetic variation and human individuality, and the relationship between linear sequence and three dimensional structure. This might sound ambitious but our participants found the concepts involved reasonably straightforward, and they all finished the activity in time. They were surprised that they could access the human genome without restriction and pleased that they'd be able to explore it further by themselves. Some participants expressed disappointment that the practical didn't involve an activity in an experimental lab, but some of them went on to do another practical as part of their tour of EMBL. The treasure hunt has since been used elsewhere: we successfully adapted it for inclusion in our CEEBT teachers' workshop (see part 3 below), and one of our volunteers is now using the treasure hunt to teach basic bioinformatics skills to non-biologists at the British Antarctic Survey.

Copies of the workbook can be downloaded from:
www.ebi.ac.uk/~schlitt/Schatzsuche/ (German version) or
www.ebi.ac.uk/~cath/treasurehunt.pdf (English version)

## 2. Wellcome Trust Genome Campus DNA day

This open day (actually two days), held in July 2003, celebrated the completion of the human genome sequence, the 50$^{th}$ anniversary of Watson and Crick's discovery of the structure of the double helix, and 10 years of the Sanger Institute. The bulk of the organization was done by the Sanger Institute, but with contributions from the other two institutes that it shares the campus with – the EBI and the Human Genome Mapping Project Resource Centre (HGMP-RC). We opened our doors to people from the local villages, secondary schools and primary schools throughout the UK, staff and their families, as well as holding a reception for eminent scientists and representatives from funding bodies.

The centerpiece of the open day was an exhibition in a large marquee in the grounds. Visitors could take themed routes through 19 different displays, exploring topics such as modeling the double helix, codes and computers, the origin of life, databanks, individuality, and ethics. There were different activities and competitions at each stand, all designed to engage the participants rather than just having them absorb information.

Practical activities ranged from a simple ethanol precipitation of DNA (done in a few moments at one of the stands) to molecular diagnostics and histopathology workbenches that took about an hour. There were also regular tours of the Sanger Institute's Sequencing Centre.

Over the two days we had more than a thousand visitors of all ages, which included school parties from all over the UK. The open day was a resounding success and the Sanger Institute plans to use it as the basis for a permanent exhibition on campus. More information about DNA day is available at www.sanger.ac.uk/Info/openday/.

## 3. What's all the fuss about genomes?

'What's all the Fuss about Genomes?', organized by the three institutes on the Wellcome Trust Genome Campus, was one of a series of nine workshops held throughout Europe during 2003 and 2004 as part of a project known as CEEBT – Continuing Education for European Biology Teachers (see www.ceebt.embo.org/ for more information). CEEBT is coordinated by the European Molecular Biology Organization (www.embo.org) and its sister organization, the European Molecular Biology Laboratory (www.embl.org), and funded by the Commission of the European Union under its 5$^{th}$ Framework Programme. CEEBT teachers' workshops are aimed primarily at secondary school teachers, and especially those teaching 16–18-year-olds. These workshops bring teachers and scientists together for two intensive days of practical classes, lectures and discussions. In so doing, they provide a forum for the exchange of best practice, information and experiences.

Before planning the workshop in detail, we held a brainstorming session with local teachers. This highlighted the need to make as much of the workshop as possible directly transferable to the classroom. Genomics and molecular biology form a small but important part of the biology syllabus for 16-18-year-olds in many European countries; our research revealed that, with the best will in the world, teachers do not have much time in which to teach these topics; preparation time and funds to buy kits and equipment are at a premium. We therefore set out to focus the majority of the weekend on practical exercises that can be done in the classroom with relatively inexpensive materials.

**Programme**

We had a complex and ambitious programme; space constraints dictated that we ran parallel sessions. After a brief welcome talk and safety briefing, the delegates were split into four groups (labelled A, T, G and C). Each group rotated around two lab-based practicals, a computer-based practical, a DNA modelling practical, two discussion sessions, a tour of the sequencing centre and a small number of keynote talks. There was also an exhibition and the exhibitors gave brief demos during coffee and lunch breaks. The bioinformatics practical was written and run by campus staff; all the other practicals were written and run by staff from the UK's National Centre for Biotechnology Education, with some help from campus-based volunteers. Each activity had a defined set of objectives that were directly relevant to biology topics taught to 16–18-year olds. For example, one of the practicals provides conclusive evidence that DNA really is the genetic material and shows that DNA encodes protein; the bioinformatics practical (exploring genomes) deliberately used examples of genes and diseases that form part of the UK's A-level syllabus, and provided an alternative means of teaching these parts of the syllabus.

Participants also had the opportunity to experience for themselves the vast scale of the genome sequencing facilities at the Wellcome Trust Sanger Institute. They familiarized themselves with how genomic data are managed and made available through the EBI's data resources, and they used some of the tools available to mine and analyse the data. Perhaps most importantly, they had an opportunity to meet some of the scientists who work at the three institutes on campus and to discuss their science with them. These activities complemented the more tightly defined, objective-driven practicals, by providing the delegates with experiences that they would be able to draw on in class.

**Lessons learned**

We were overwhelmed by the positive responses from the delegates; feedback indicated that the attendees left the campus enthused and raring to transfer this enthusiasm to their students. Over 70% of the delegates completed the feedback form and all said that the workshop either lived up to or exceeded their expectations.

Many of the delegates were delighted (and some were quite emotional) about being able to visit the bosom of genomic sequencing in Europe; they praised the provision of resources that really could be taken straight into class, the clarity of the presentations and the willingness of speakers to answer questions.

One of the greatest difficulties that we faced when organizing the workshop was its effective promotion. However, the process of organizing this workshop has provided us with outlets for publicizing future ones. A more worrying concern is that, although initially we had over 86 registrants, only 56 teachers actually attended. For future workshops we might have to charge a deposit, refundable on attendance, to make sure that we don't waste valuable places that could have been taken up by enthusiastic teachers.

A full report of the workshop, which includes all the practical protocols and slides from the lectures, is available from www.ceebt.embo.org/countrydescriptions/hinxton.html. Similar reports from the other CEEBT workshops held throughout Europe are available from www.ceebt.embo.org/ceebttw.html.

## 4. Concluding remarks

What are the ingredients for a successful outreach event? Each event had its positives and negatives (Table 1), but we've learned several lessons during the planning of these three events that are relevant to any outreach activity.

**Table 1.** Pros and cons of each outreach activity.

| Activity | Pros | Cons |
|---|---|---|
| Exploring your genome | • Cheap and simple to run<br>• Effective way of teaching both biology and IT skills<br>• Participants can do the activity at home or in class if they have access to the Internet | • More effective when coupled with an experimental practical, which drives up the cost<br>• Limited to a small number of participants or many repeat runs of the same activity |
| Genome Campus DNA day | • Effective way of reaching many people: overall budget is high but per-capita costs are low<br>• Marquee allows participants to explore many aspects of the topic at their own pace<br>• Involved the entire campus in the public communication of science | • Needs a large budget (~€75 000)<br>• Very time consuming to organize<br>• Not practicable for many academic institutes |
| What's all the fuss about genomes? | • By 'training the trainers', has potential to reach many young people<br>• Comprehensive<br>• Carefully targetted to a specific audience | • Needs a significant budget (~€25 000 for 80–100 participants); per capita costs are high<br>• Requires significant investment of time for participants<br>• Limited to a small number of participants |

**Outreach is a two-way process**

Planning any type of outreach event involves educating the scientists who'll be taking part in the event, as well as educating the participants. The volunteers need to be well briefed about the purpose of the event, its target audience, and what can realistically be done with the time/space/money available. It helps to lay down some ground rules and stick to them (for example, for the Genome Campus DNA day we limited the amount of information that could be put on posters in the exhibition, to make them as readable and clear as possible). If your institute hasn't run an activity of this nature before, seek advice from educational experts from the very beginning. For example, a couple of hours spent discussing our CEEBT teachers' workshop with staff at the National Centre for Biotechnology Education (Table 2) made us rethink our plans, and made the entire event much more relevant to our target audience.

**Do some market research**

This is especially important if you have a tightly defined target audience. For example, to help us organize our teachers' workshop we had a brainstorming session with potential delegates. This helped us to think about the practicalities (e.g. when to hold the workshop) as well as the science (e.g. how much molecular biology and genomics do 16–18-year olds cover? How can we make the contents of our workshop 100% relevant to their syllabus?).

**Do a 'dry run' if possible**

For example, we tested our genome treasurehunt on the children of some of our staff members to make sure that we'd pitched it correctly and that it wouldn't take longer than the allocated time.

**Keep it simple**

You do not necessarily have to go high-tech to generate enthusiasm for your subject. For example, a simple DNA precipitation that took only a few moments was an extremely effective way of bringing practical science to a large number of people at the Genome Campus DNA day.

**Add as much realism as possible**

The DNA precipitation gave participants a chance to step into the shoes of real scientists. They used the same equipment as a bench biologist would use (Eppendorf tubes, Gilson pipettes) and they performed a task that is an integral part of modern molecular biology. Similarly in our bioinformatics practicals, participants used the same data and tools as biologists use. It's this authenticity that they find thrilling.

**Promoting your activity is VERY important**

People won't come unless they know about it! Plan your promotional campaign right at the beginning and don't expect more than 5–10% of the people you reach to attend the event.

## Seek feedback

It's difficult, if not impossible, to gauge the long-term success of an outreach event; how do you measure whether your event influenced somone to go into a career in science? But you can at least find out whether your event was stimulating and enjoyable for the participants. If you're planning to run regular events you'll need to find out what didn't work well for the participants, so that you can put it right for next time. Feedback can also be a great motivator for the organizers and volunteers involved in outreach projects. Think carefully about what you need feedback on, incorporate appropriate questions into your evaluation form and, if necessary, provide incentives for participants to return their evaluation forms.

**Table 2.** A few useful websites.

| URL | Description |
|---|---|
| www.ebi.ac.uk/2can | 2can Bioinformatics: the EBI's educational website |
| www.yourgenome.org | A user-friendly guide to genomics |
| Openday websites | An extensive list of educational resources collected for our Campus Openday held in July 2003 |
| www.wellcome.ac.uk/genome | The Wellcome Trust's guide to the human genome |
| www.biology4all.com/ | Useful information for pupils and teachers, including a database of teaching resources |
| www.biology.arizona.edu/ | The University of Arizona's online interactive resource for learning biology |
| www.ncbe.reading.ac.uk | Protocols and equipment for teaching biotechnology in schools, developed at the University of Reading |
| www.ceebt.embo.org/ | This initiative, funded by the European Commission and organized by EMBO, runs workshops and labs for teachers as well as providing online resources |
| www.dnalc.org/ | Educational links, protocols, workshops and more from Cold Spring Harbor Laboratory. |
| www.ase.org.uk | The UK's professional association for teachers of science |
| www.sciencelearningcentres.org.uk | The UK's national network of centres for the professional development of science teachers and technicians |

**Acknowledgements**

I am indebted to the following for their support and enthusiasm: EMBL's Office of Information and Public Affairs, for involving us in EMBL open house 2003; many colleagues at the Sanger Institute, especially Christine Rees, Alison Coffey and David Bentley, for the opportunity to take part in the Genome Campus DNA day; EMBO (especially Andrew Moore and Sandra Bendiscoli) and the European Commission for making the CEEBT workshops possible, as well as my co-organizers, Don Powell (Sanger Institute) and Lisa Mullan (then at the HGMP-RC, now at the EBI); The National Centre for Biotechnology Education, especially Andy Harrison, John Schollar and Dean Madden, for masses of advice and boundless enthusiasm; and my many colleagues at EBI for helping to plan, devise and run the activities discussed. I would especially like to thank Thomas Schlitt, Manuela Pruess, Esther Schmidt, Peter Rice and Rolf Apweiler for devising, translating and running 'Exploring your genome: a treasure hunt on the Internet', and a list of people too numerous to mention individually for their involvement in the Genome Campus DNA day. Last, but definitely not least, I thank all the participants in our activities for their enthusiasm and helpful feedback.

# The English Patient: Biotechnology Education in the UK

**Dean MADDEN**
*NCBE, The University of Reading,*
*Reading, RG6 6AP, United Kingdom*
*D.R.Madden@reading.ac.uk*

**Abstract**. In the mid-1980s, the UK government established several initiatives to encourage and enhance biotechnology education in schools. 20 years later, only one project, the National Centre for Biotechnology Education (NCBE), remains. It has however, an international reputation and reach. What lessons can be learnt from the English experience, and could it provide a model for other initiatives?

## Introduction

There is a good old tradition (as good and as old, in fact, as Plato and Aristotle) which demands that those who tackle a problem and propose a new solution should first give a critical account of its history. The majority of this paper is therefore a historical review of the development of biotechnology education in England and Wales. It sketches a decline in the coverage of biotechnology in the school curriculum from the mid 1980s, when the UK was (albeit very briefly) an acknowledged world leader, to the present.

## 1. Biotech means business

Biotechnology entered the school curriculum (principally the *biology* curriculum) in England and Wales during the early 1980s. Three influential reports stimulated this. In 1980, a report from a joint working group of the government-funded Research Councils (commonly known as the Spinks' Report [1]) defined biotechnology as *"... the application of biological organisms, systems or processes to manufacturing and service industries"*. This broad definition, encompassing both traditional and 'new' biotechnologies, was also adopted the following year by a Royal Society review of biotechnology education [2]. The Royal Society's report encouraged teachers to include relatively simple practical work (such as food fermentations and basic microbiology) in their courses. The UK's approach contrasts with that initially taken in other countries, such as the United States of America [3] and Denmark [4], where biotechnology was regarded as synonymous with recombinant DNA technology. Genetic modification was rarely mentioned in a school context in the Britain at this time [5], essentially because it lay beyond the knowledge and experience of all but the most recent graduates and was thought to be too dangerous or costly to be studied practically in schools [6].

In 1984, a UK Department of Trade and Industry (DTI) survey of the awareness of biotechnology among school teachers [7] concluded that while many biology teachers appreciated its economic importance, there was little educational incentive for them to incorporate biotechnology into their teaching, because the topic was seldom mentioned in examination specifications. Furthermore, few teachers had the appropriate training or experience to do this.

## 2. The National Centre for School Biotechnology

John Grainger at The University of Reading was already well aware of the problem highlighted by the DTI research. In 1983 he had persuaded the Society for General Microbiology (SGM) to fund a project in the Department of Microbiology at the university to further the teaching of practical biotechnology and microbiology in schools. The resulting book and collection of 30 photocopyable workcards, *Practical microbiology and biotechnology for schools* [8], proved popular and on publication it was referred to by most post-16 biology syllabuses, although the pack was aimed at younger students.

In September 1984, the DTI followed the SGM's lead by establishing (under its first director, Paul Wymer, who was the author of the SGM publication) the National Centre for School Biotechnology (NCSB), also in the Department of Microbiology at Reading. Nobel Laureate James Watson, who at the time was involved in trying to set up a similar 'DNA Learning Center' at his Cold Spring Harbor Laboratory, remarked: *"Your National Centre for School Biotechnology represents the first national commitment to update pre-university teaching to take into account the dramatic rise of DNA science"*.

The Centre's main role was to encourage and support the teaching of biotechnology in schools and colleges. To achieve this, an extensive library of educational resources (books, video recordings, 35 mm transparencies, computer programmes, *etc.*) was built up. Teachers could visit the library or telephone the NCSB to ask for help and advice. Several computer databases were developed so that the Centre could put teachers with an interest in biotechnology in contact with one another (at the time this was fairly revolutionary as desktop computers had only just been introduced). The Centre also published a newsletter which was distributed free-of-charge to all secondary schools in Britain and to many people overseas. Thus, to begin with, the Centre's activities focused on bringing together like-minded enthusiasts and informing teachers about the educational resources that were available to help them.

In recognition of the importance of in-service teacher training in biotechnology, the Manpower Services Commission (then a branch of the UK Department of Employment with particular responsibility for training) funded the secondment of a teacher (John Schollar) to the NCSB for three years from 1986. His task was to develop a package of in-service training materials for teachers, and to organise and run practical training workshops on location in school laboratories throughout the UK. Other DTI-funded projects followed at the universities of Sheffield, Surrey, and at the (then) South Bank Polytechnic in London, with the NCSB loosely co-ordinating the four initiatives.

## 3. The University of Sheffield

Several teachers were seconded to the Division of Education at the University of Sheffield for three years. This project studied teaching methodologies and paid particular attention to how the seconded teachers' experience affected their classroom practice, using the teaching of controversial issues as a vehicle for studying this. Surprisingly, biotechnology at the time presented few concerns, and the group resorted to tackling fears about food additives that were then widespread [9]. This reflects what little consensus there was about the definition of biotechnology within the education community at the time. The Sheffield project was more interested in educational research into continuing professional development for teachers rather than in biotechnology education *per se* and lacked a relevant scientific input.

## 4. The University of Surrey, Guildford

Technology buses became popular during the mid-1980s, particularly as few schools were then equipped with computers and other resources needed for teaching technology. A 'biotechnology bus' was based at the University of Surrey, and in-service courses for teachers were run throughout southern England by ex-biology teacher Anne Riggs. The bus was not purpose-built, but the interior was converted to form a small laboratory for 10–12 teachers. With little support from elsewhere, Anne devoted considerable enthusiasm and dedication to the project, but its impact was limited. Firstly, the bus was old and required frequent repairs: compared with the purpose-built 'computer buses' its image was hardly high-tech. The overheads of the project were also comparatively high. Secondly, the space on board was cramped, and provided little or no advantage over working in a normal school lab alongside which the bus was often parked. It may also inadvertently have given the impression (although not to teachers attending courses) that special facilities were required before practical biotechnology could be attempted in schools.

## 5. South Bank Polytechnic, London

The London project, based at the (then) Polytechnic of the South Bank, but officially hosted by The London Centre for Biotechnology (a consortium of the polytechnics of Central London, The South Bank and Thames) ran several residential in-service training courses for teachers from across Britain between 1986–8. These were well-received by the participants, who were often engaged in a major DTI-funded curriculum development (the Technical and

Vocational Education Initiative, TVEI). Initially, the courses centred on traditional applied microbiology, at a slightly higher level than might usually be encountered in a school. Subsequent courses covered a broad range of biotechnology, introducing, for example, a microbial fuel cell which has remained popular in schools ever since. These courses, run by Irena Oljenik and the staff of the polytechnic, genuinely inspired many teachers. The project had, however, no means of disseminating its good practice further afield or, crucially, of supplying schools with the specialist materials that were necessary for much of the practical work to be replicated in the classroom. A few years later the consortium that formed The London Centre for Biotechnology was dissolved.

## 6. Curriculum innovation

These developments were taking place against a backdrop of substantial curriculum innovation, through the implementation of the TVEI in some regions of the UK, the introduction of novel methods of formative and summative assessment (such as records of achievement to supplement formal examinations) and the implementation of early internet-based support services (such as NERIS, the National Educational Resource Information Service).

The arrival of the General Certificate of Secondary Education (GCSE, generally studied between the ages of 14 and 16, which replaced the former two-tier system of 'O' Levels and CSE examinations) in 1988 afforded an opportunity for wider, more mainstream curriculum innovation, although the earlier '16+' examinations which had effectively been pilot studies for the GCSE had also incorporated aspects of biotechnology. Several GCSE courses of the late 1980s featured considerably more biotechnology than is found at present. For example: a Modular Science GCSE included an eight-week 'Enzyme Technology' module; the 'Suffolk Science' GCSE included compulsory practical plant tissue culture; and the Warwick Process Science Project had a substantial biotechnology module. Several local education authorities, often under the auspices of the Secondary Science Curriculum Review and the TVEI, developed courses and published biotechnology resources, notably in Staffordshire, Sheffield and Strathclyde (these Scottish materials were seized upon and used by teachers throughout the UK) [10, 11]. At the University of Newcastle-upon-Tyne, biology teacher Dean Madden was developing resources for Cumbria's TVEI programme, and contributing to the development of NERIS as well as several other initiatives.

From our current perspective it is hard to appreciate the scale and range of the innovation undertaken by teachers at the time. Even the Association for Science Education had a Biotechnology Working Party, which published resources and advice for schools [5, 12]. There was little agreement, however, between education professionals about the extent to which biotechnology should feature in the secondary school curriculum, if at all. 'Biotechnology or basics?' was a typical headline in the educational press [13].

To try to achieve a consensus, the MSC funded a national conference at the University of Kent at Canterbury in 1986. It was greatly oversubscribed and attracted 280 delegates, bringing together teachers, local education authority science advisors and representatives from industry and universities. Similar meetings had previously been held in Cardiff and London. One of the architects of the recently-proposed National Curriculum, Professor Paul Black from King's College London, warned the Kent meeting that school biotechnology needed to define itself more precisely if its place in the curriculum was to be assured, and particularly if it was to be capable of being assessed for examination [14]. Sadly, although some understood his words, few appreciated their prescience and fewer still were in a position to influence the development of the National Curriculum.

As the '80s drew to a close, the DTI's support for education was rapidly wound down and the Department for Education and Science (DES) reasserted its authority over schools and the curriculum.

## 7. The National Curriculum

In 1989, the UK government imposed the National Curriculum in England and Wales, dictating what was to be taught to every child between the ages of 5 and 16 years. The death thoes of the old system were prolonged however, as the introduction of the National Curriculum was phased, with the first examinations at Key Stage 4 (that is, at the age of 16) taking place in 1994. The first version of the National Curriculum [15] incorporated many aspects of biotechnology. For example, students had to:

- understand the role of microbes in sewage disposal and composting;
- be able to make informed judgements about the benefits and drawbacks underlying the use of hormones and growth regulators in controlling growth, development and fertility in plants and animals;
- have opportunities to explore and investigate the uses of enzymes and microbes, for example, in the brewing, baking and dairy industries;
- be able to describe a range of cloning methods and their uses in agriculture, and show an appreciation of the economic impact and related ethical issues;
- understand the basic principles of genetic engineering in relation to drug and hormone production;
- be able to make informed judgements about the economic, social and ethical issues concerning the recent developments in genetic engineering;
- understand the application of biochemical processes in manufacture, for example, in fermentation.

This list is not exhaustive. However, compared with some GCSE Science courses, the National Curriculum represented a considerable *reduction* in the explicit coverage of biotechnology. In addition to introducing a legally-binding curriculum, the government made other significant changes to the ways in which schools were run, including local financial management of schools, inspection of schools by private companies and the publication of examination league tables. One consequence was a reduction in the influence of local education authorities (LEAs) and of Her Majesty's Inspectorate of Schools (HMI) in England and Wales. In the new circumstances, co-operative curriculum development (which had often been fostered by LEA science advisors and promoted by HMI) became a rarity. Schools were now in effect businesses competing with one another. The focus shifted increasingly towards the specific content that was demanded by law and more importantly, the associated national tests. Virtually all of the innovation and debate about biotechnology education ceased.

## 8. The wilderness years

As the DES asserted its authority, the DTI withdrew entirely from the school sector. Products of gene technology (*e.g.*, enzymes and drugs from GMOs) began to reach the market, and nervous companies were increasingly reluctant to bring their products to the attention of the public. Opposition to biotechnology was developing. In 1989, the employers' organisation, the Confederation of British Industry (CBI) proposed restrictions on public access to information on releases of GMOs into the environment *"... until understanding is more widespread"*.

A Greenpeace consultant, writing in the Summer 1987 issue of the NCSB newsletter, had rejected the idea of *"... insensitive or inaccurate opposition to the whole phenomenon"*, suggesting instead that *"we must apply ourselves to the task of educating and informing about this new technology, with neither panic nor blind faith in progress."* A few years later attitudes had changed, particularly those of Greenpeace. DTI support for the various biotechnology education initiatives ended in 1990 or shortly after. The NCSB's staff, now joined by Dean Madden, made plans to become self-funding by charging for the newsletter and courses. The Centre also changed its name to the National Centre for Biotechnology Education (NCBE), with the aim of providing services to a wider range of clients such as businesses and members of the public. None of the other DTI-funded projects were able to make similar arrangements, and all of the other biotechnology education initiatives therefore ended.

In 1990, following advice from a DTI business consultant, the NCBE established a 'Biotechnology Club'. Schools joining the Club received a teachers' newsletter and discounts on the limited range of products supplied by the Centre (such as small quantities of industrial enzymes). Unfortunately, the lower circulation of the subscription newsletter curtailed the income from advertising and opportunities for sponsorship dried up as recession bit. Economic recession in the early- to mid-1990s contributed to many biotech start-up companies either ceasing to trade or being taken over by larger concerns. Businesses within the sector became less accessible to schools and consequently the pool of resources (both financial and informational) that schools could draw upon diminished. Things looked bleak and in 1991 the Centre's director departed for another job. John Grainger nominally took over as Director, but the Centre was now run by the two Assistant directors, John Schollar and Dean Madden.

The NCBE's 'Biotechnology Club' continued for two years but eventually the cost of administering the collection of thousands of cheques from individual schools made continued production uneconomic. Similarly, Pergamon Press was unable to achieve a high circulation for the academic journal *Biotechnology Education*, which had been started by and was still edited by the now-departed Paul Wymer. The title was sold to another publisher, but it did not prove a success. Like the NCBE's newsletter, the journal ceased publication in 1992.

A major consequence of the National Curriculum and other changes in education was to stifle creativity and innovation. Because few teachers would risk going beyond the bounds of what it was legally required of them to teach, few publishers produced books covering material that was not required by the curriculum. Unsurprisingly, the regular stream of visitors to the NCBE's 'resources room' slowed to a trickle then ceased altogether. Innovation by British educational suppliers was also restricted. In 1990 there were two main science education suppliers in the UK: Griffin and George and Philip Harris Education. The early 1980s had seen the introduction by both firms of numerous simple 'biotechnology kits'. Philip Harris's were all developed in-house, while Griffin also imported kits, unmodified, from the USA. By 1983–4, there were about 40 different biotechnology 'kits' on the UK market, and both of the main suppliers produced specialist biotechnology brochures. But by 1992 Philip Harris had closed its biological sciences research and development subsidiary (Philip Harris Biological) and shortly afterwards Griffin shut down a large part of its biology-related business.

By now, the NCBE's remaining staff was on monthly contracts and some had left. It looked as if a combination of economic recession, growing opposition to gene technology and an inflexible curriculum would kill biotechnology education in the UK.

## 9. Courses and sales

Although there are no specific qualifications in biotechnology in secondary education in England and Wales, almost all post-16 ('A' Level) school biology courses provide opportunities for students to study aspects of biotechnology. By the early 1990s the (then) eight 'A' Level examination boards in England and Wales covered a broad range of biotechnology [16]. This content was mainly in the optional sections of biology and human/social biology syllabuses so it was not necessarily covered by all students. Biotechnology and microbiology options were, however, the most popular ones on offer with more students and teachers selecting these for study than all the other options combined. For example, when the first cohort of 'A' Level Biology students sat the (then) new University of London examinations in June 1992, 56% chose to answer questions on the Microbiology and Biotechnology option, compared with 13% and 31% for the other two topics on offer.

In partnership with the far-sighted biology secretary, Erica Clark, at the London examination board and subsequently with other organisations, the NCBE ran numerous teacher training workshops, which not only provided a direct income, but also helped to boost the sales of the materials that the Centre provided. In 1992, this was restricted to a limited range of enzymes, a simple 'bioreactor' or fermenter and copies of Paul Wymer's *Practical biotechnology and microbiology* workcards. By September 1993 these were joined by inexpensive DNA gel electrophoresis equipment and a *Practical biotechnology for schools* booklet [17] that had been devised by the Centre's Assistant directors. Throughout the 1990s, industrial and government support for the Centre was almost non-existent but by the mid-1990s it was obtaining a regular income from training courses for teachers and increasingly, post-16 students, plus sales of equipment and materials. No doubt this was aided by the vacuum created by the closure of biology research and development by the UK's main school suppliers.

Another factor that contributed to the Centre's survival was its decision not to create a prestigious laboratory and visitor centre with the associated overheads that such a facility would incur, but to focus instead on running courses on location, in school laboratories and at other venues. In part this policy was determined by financial constraints and the lack of sponsorship available in the UK, but it was also informed by the experience of the 'biotechnology bus' at Guildford and observation of similar projects elsewhere.

## 10. The NCBE today

Against the odds, the Centre has continued to prosper and is well-known to schools in the UK and further afield. In contrast to many other initiatives throughout Europe and the USA, the NCBE's aim has always been to bring biotechnology to as many students as possible, by fostering capability amongst teachers and providing low-cost training, equipment and materials.

It has gained an international reputation for the development of innovative educational resources: its materials have been translated into many languages including German, Swedish, French, Dutch and Danish. The NCBE was a leading founder member of EIBE, the 'European Initiative for Biotechnology Education' (www.eibe.info), an EC 'Concerted Action' which ran until 2000. A national poll conducted for The Wellcome Trust in 1999 showed that biology teachers considered the NCBE the best science education organisation in the United Kingdom.

NCBE equipment is sold to more than 30 countries world-wide, including North America, where several items are licensed to a major educational supplier (Carolina Biological). The NCBE remains, however, part of The University of Reading and a not-for-profit organisation. The Centre's patented equipment for DNA gel electrophoresis was granted 'Millennium Product' status by the UK Design Council and was one of three biotech-related items to be taken on a prestigious Science Museum tour of Japan (the others being Watson and Crick's DNA model and a sweater made from the wool of *Dolly* the cloned sheep). Independent research has shown that through the provision of materials and training courses, the NCBE is successful in ensuring that these new procedures are adopted by teachers in the classroom: over 500,000 sets of NCBE electrophoresis equipment have now been sold world-wide.

The NCBE's Web site (www.ncbe.reading.ac.uk) was started in January 1995 and is recognised as a valuable source of information. It has featured in *Nature Biotechnology* and attracts more than 100,000 connections per week. The Centre also deals with written and telephone enquiries from teachers, students and members of the public each day, particularly on safety and practical project work.

The NCBE has become Europe's principal provider of in-service training for school biotechnology and has run courses in eight EU member states, mainly for teachers, student teachers and post-16 students. On average, more than two courses a week are run by the Centre during term-time (about 4,000 people attend NCBE courses per year). For the last 11 years, the Centre has run residential summer schools in Sweden, organised in association with and accredited by the University of Göteborg.

## 11. New challenges

Curriculum change has continued unrelentingly throughout the 1990s and to the present day. With major revisions in 1995 and 1999 [18] the content specified by the National Curriculum was further reduced. The only elements of biotechnology mentioned at Key Stage 4 (ages 14–16) in 2004 are:

- the basic principles of cloning, selective breeding and genetic engineering;
- how enzymes may be used in biotechnology.

These most recent changes have been made as the shortcomings of a rigid, centrally-controlled science curriculum have become increasingly apparent over the last decade [19]. It is hoped that from September 2004, the limited requirements at Key Stage 4 will permit greater flexibility and possibly the inclusion of novel content once more.

In September 2000, the introduction of a new A/S-A Level structure significantly reduced the opportunities for studying aspects of biotechnology at that level, compared with the content specified in the previous 'A' Level modules. Several topics have been lost completely, such as ELISA and other aspects of immunology. Discrete 'modules' of work in the new post-16 courses have contributed to a lack of coherence in the way some topics are dealt with, particularly inter-related ones such as biochemistry, genetics and microbiology. The modular structure also makes it difficult for students to carry out longer-term practical investigations, which were a characteristic of some courses under the older system. These shortcomings and many others of the new post-16 examinations are now widely-recognised and a thorough review has been ordered by the UK government under Mike Tomlinson, a former head of the Office for Standards in Education (OFSTED).

Although some preliminary proposals have been aired it is not currently known (in September 2004) what will replace the current A/S-A Level arrangement. In England, the number of examination boards (now called 'awarding bodies') has shrunk from eight (in 1990) to three (in 2004) and they have all become businesses, loosing their former affiliations with universities. Chief examiners (who devise the courses) now tend to be drawn from the ranks of older and more experienced teachers, which has lead to perhaps excessive concentration on well-established topics and only cursory treatment of modern biology, even though its importance is recognised by teachers and their students [20]. For example, the human genome project receives a single line if it's mentioned at all in the current A/S-A Level specifications, and bioinformatics is nowhere to be seen. These and several other factors have stifled curriculum innovation post-16, and have no doubt contributed to the decreasing popularity of science studies that seems to be an international phenomenon.

With this in mind, in October 2004 a new £51 million national network of nine Science Learning Centres for teachers will open in England, followed by a larger National Centre in 2005 (www.sciencelearningcentres.org.uk). Their sponsors, the government's Department for Education and Skills and The Wellcome Trust, hope that the new centres will inject new life into science education in the UK. The NCBE will be involved in offering courses at these venues, and will continue to help teachers provide innovative, up-to-date and stimulating bioscience education for their students.

**References:**

1. Advisory Council for Applied Research and Development / Advisory Board for the Research Councils / The Royal Society, Report of a Joint Working Party. Biotechnology. Her Majesty's Stationery Office, London, 1980.
2. The Royal Society, Biotechnology and education — The report of a working group. The Royal Society, London, 1981.
3. D. Micklos and G. Freyer, DNA Science — A first course in recombinant DNA technology. Carolina Biological Supply Company / Cold Spring Harbor Laboratory Press, New York, 1990.
4. H. Agensen *et al.* Experimental gene technology in education. English edition. Nucleus Forlag ApS, Århus C, 1992.
5. Association for Science Education, Biotechnology Sub-committee, Biotechnology in science examination syllabuses, *Education in Science* **111** (1985) 16–19.
6. HMSO, Microbiology. An HMI guide for schools and further education. Her Majesty's Stationery Office, London, 1985.
7. J. Beetlestone and J. Teasdale, Awareness of biotechnology in schools. Department of Trade and Industry, London, 1984.
8. P.E.O. Wymer and J. Grainger, Practical microbiology and biotechnology for schools. Macdonald, London, 1986.
9. J. Henderson and S. Knutton, Biotechnology in schools — A handbook for teachers. Open University Press, Milton Keynes, 1990.
10. R.E. Phipps, Biotechnology in schools, *Education in Science* **120** (1987) 21–22.
11. B. Godding, The development of biotechnology in Staffordshire schools *Journal of Biological Education* **22** (1988) 211–214.
12. J. Richardson, Practical work with DNA, *Education in Science* **162** (1985) 16–18.
13. J. Dunkerton, Biotechnology in schools — Innovation or stasis? *School Science Review* **68** (1987) 489–492.
14. P. Black, Biotechnology — a new technology in education? In: Biotechnology within the curriculum. A TVEI / TRIST perspective. Report on a conference held on the 4–6 July 1986 at the University of Kent at Canterbury.
15. Department for Education and Science, Science in the National Curriculum. Her Majesty's Stationery Office, London, 1989.
16. D. Madden, Biotechnology in 'A' Level biological science syllabuses, *NCSB Newsletter* **8** (1990) 3.
17. D. Madden, Practical biotechnology — A guide for schools and colleges. NCBE, Reading, 1993. (Available from: www.ncbe.reading.ac.uk)
18. Department for Education and Employment, The National Curriculum — Handbook for secondary teachers in England. The Stationery Office, London, 1999. (Available from: www.nc.uk.net).
19. House of Commons Select Committee on Science and Technology, Third Report: Science education from 14 to 19. The Stationery Office, London, 2003. (Available from: http://www.parliament.the-stationery-office.co.uk/pa/cm200102/cmselect/cmsctech/508/50802.htm)
20. The Wellcome Trust, Life study — Biology A level in the 21st century. The Wellcome Trust, London, 2004. (Available from: www.wellcome.ac.uk)

# Session V

# Successful Practices for Research Training – Central-Eastern Europe

*Science Education: Best Practices of Research Training for Students under 21*
*P. Csermely et al. (Eds.)*
*IOS Press, 2005*



# Weekend Scientific Seminars and Lectures for High School Students

**Gagik SHMAVONYAN, Lili KARAPETYAN,**
**Gayane SHMAVONYAN and Nelly YEGHIAZARYAN**
*Armenia-Great Britain Society, Yerevan, Armenia,*
*gshmavon@yahoo.com*

**Abstract**. The results of researching on few hundred high-school students of Armenia, mostly from the capital of Armenia, Yerevan for one year showed us that the weekend scientific seminars and lectures are very important and helpful for high school students, which help them to choose their future profession and finally promote science. The aim of the weekend seminars and lectures is to explain to high-school students and their parents the role of the science, its application in our life, to understand differences in scientific fields, to help in choosing the scientific field in which they are more interested, help them understand many popular scientific topics, and to introduce the latest scientific, technological achievements and news to them.

## Introduction

Armenia-Great Britain Society is a non-governmental Organization. Its members are University Professors, teachers, students, schoolchildren and industry professionals, etc. One year ago the members of our Society, who are representatives of higher educational institutions and people engaged in Industry, noticed that high-school students, undergraduate and graduate University students have problems in choosing their further profession, especially in different fields of science. The reason is that many of them have not sufficient information on the real differences between the fields of science, which field of science their scope of interests includes.

Promotion of science is actual in our century. Many high-school students, their parents and graduate students, who want to continue their study further, are not oriented how to choose the further profession and they need advice from renowned scientists from Universities, academies of sciences, research institutions and industry specialists.

To solve this problem the members of our Society decided to organize weekend scientific seminars and lectures for interested people.

## 1. Definition of the Problem

According to our investigation during one year we would like to find out what percentage of high-school students, especially gifted and talented ones are oriented in choosing their profession, especially in different fields of science.

To promote the science we decided to organize weekend scientific seminars and lectures for the interested people, especially for the high-school students. As the lectures are mostly held by University Professors and professionals of industry, who usually are busy in working days that's why we decided to organize the scientific seminars and lectures on general scientific topics at weekends. Mostly the weekend scientific seminars and lectures are held on Saturdays and occasionally on Sundays.

## 2. Results of Research Practice

The Board of Society is organizing weekend scientific seminars and lectures on general scientific topics, especially those which are of interest to the high schoolchildren, as well as scientific and educational tours to the laboratories of different Universities. Every weekend a scientific topic of weekend seminar and lecture is chosen by the Board of the Society and one of the Professors holds a benevolent lecture of nearly one hour duration. After the lecture the discussion of the scientific topic of the lecture or seminar takes place.

Then two or a few groups of interested people, high-school students and students attend University laboratories. The duration of University laboratory tours is about 20 minutes. During laboratory tours University Professors or PhD students in popular way introduce to them the scientific equipments and tools of laboratory and the experiments carried out in that laboratory. These weekend scientific lectures and seminars, which are associated mostly with pictures, help high-school students and graduate students to get general information on different fields of science, to choose their further profession and to learn the latest scientific and technological news.

Such kind of weekend scientific seminars and lectures allow the interested people, especially high schoolchildren to get information on different fields of science and finally to choose the appropriate field of science after leaving school and further graduating from higher educational establishments. It is very important to organize weekend popular seminars and lectures with beautiful illustrations and organize laboratory demonstration – scientific and technological equipments, etc. It is also very useful to organize scientific tours to science parks or science museums, which will allow interested students to get comprehensive information on different fields of science and increase their knowledge of science, to get acquainted with scientific and technological achievements.

## 3. Summary and conclusion

To reach better quality in education and better talent recruitment means not only better material resources, but what is more important, better human and psychological circumstances – more stimulating environment for growth and development of talent: better motivition, better education of teachers and differentiated program.

We can see that depending on the scientific topic different quantity of interested people are participating in these seminars and lectures. So, the choice of the scientific topic is very important : it should be popular and should be presented and explained in a way so that students will be able to easily understand what is being spoken about. Besides, it is very important to choose the time of the lecture or seminar. We found out that 10:30 a.m. is very convenient for the most of the attenders. After discussions during the weekend scientific seminar or lecture many of the interested people continue interested in the topic also in further.

To organize such kind of weekend scientific seminars and lectures we do not need sponsors. It is important to have willigness of the organizers and lecturers to do benevolent work, to organize the seminars and lectures and help high-school students and interested people.

The questionnary which was filled by the particpants of the weekend scientific seminar and lecture allowed us to improve our lectures and seminars helping to choose further scientific topics and lecturers.

In conlusion, we noticed that in the beginning of the series of the seminars and lectures the students seemed to be more interested in ecological issues and language problems. After half year we noticed that students got interested in natural sciences as well, especially physics, astronomy and medical education.

Besides, we understood that these weekend scientific lectures and seminars are of great help in understanding the demands of high-school students, to reveal gifted and highly talented students. They also proved to be a very good way of bringing people with the same scope of interests together.

In further we are going to improve weekend scientific seminars and lectures, find ways for promoting science among high-school students.

**References:**

[1] G.A. Davis, S.B. Rimm, Education of Gifted and Talanted, Prentice Hall, Englewood Cliffs, NJ 1989.
[2] S.J. Sorenson (editor), The Gifted Program Handbook, Planning, Implementing, and Evaluating Gifted Programs, Dale Seymor Publication, Palo Alto, 1987.

*P. Csermely et al. (Eds.)*
*IOS Press, 2005*


# Talent Support in the Czech Republic from the Gifted Children's and their Parents' Point of View

**Eva VONDRAKOVA**
*STaN-ECHA,*
*Bellusova 1827/53, 155 00 Praha 5, Czech Republic*
*vondrakova@mistral.cz*

**Abstract**. This meeting aims to improve chances of the gifted students around the world to realize their potential, namely in natural sciences. We can exchange "know-how", what to do to realize our goals. Schools are obviously proud on their students´ success in competitions. These excellent students are highly intelligent, work hard and behave well. It seems to be no problems.there. Despite it some of these students are strongly dissatisfied with the way their school treats them. What are the problems connected with fulfilling gifted students´ potential? What can we do to support their motivation to intellectual growth and its realization?

## 1. Introduction

STaN-ECHA is the Czech and Slovak ECHA branch. More than 10 years there are Club for parents of Gifted Children´s (GC) meetings in Prague and seminars for teachers and psychologists, organized by Eva Vondrakova. Many parents from the whole Czech Republic look for advice and a help in their gifted children education.

STaN-ECHA started its work as early as 1989. Club for Parents works since the year 1993. From that time hundreds of parents asked our advice and help with their children´s education. Among main problems GC had at schools there were boredom, loneliness and sometimes bullying. Also underachievement, learning disabilities and behavioral problems, often because of nonconformity of the GC (mostly boys).

Many teachers and psychologists believed it was unnecessarry or even immoral to care for gifted and refused to care of them. Small number of specialists who tried to change situation in GC education was criticized by majority of laymen, as well as teachers and psychologists. Some amount of personal courage or inconcern about prevalent opinion of society was needed not to give up continuing in such work. More precisely – not work, but hobby. It was not possible to have gifted as job description. It took several years of hard work and popularization to change the attitude to the gifted in our society. Thanks to better awareness of this topic things are looking up in the last few years. More and more teachers and psychologists look for a help or cooperate with STaN-ECHA.

At the present parents are more demanding to schools than before and more active in looking for the way of their children schooling. The age of children, whose parents contact us now is lower than several years ago. At the beginning we usually helped to solve problems GC had at schools (and it continues). Newly many parents of preschool children and toddlers want to know if their child is really gifted and "how to rise a bright child" (Joan Freeman´s book title, one of books we use in our work with parents of young children). They prefer preventing problems to their solving later. Some parents are well educated and ambitious immigrants, able entrepreneurs or people who lived in abroad for several years. They know what future competenties their child has to acquire to be successful in his/her adult life and future profession no matter where he/she would be living. They usually have experience with various ways of educating children in other countries. In the Czech Republic they are looking for the schools corresponding to their ideas. If they find such school some of them do not hesitate to change their address if needed. It was not usual in our country until now.

This meeting is focused on high school students. But motivation for science begins often at an early age. As mother of Matyas told us at our Club for parents of GC meeting: When Matyas was 4 years old, his mother was telling him a story how Tyrannosaurus Rex was friend with Dimetrodont. Matyas looked at her and told: "But mum, Dimetrodont lived at Permian and Tyrannosaurus at Top Chalk!" Mother was shocked and did not understand how her son could find it out. Then she recalled that she had show him a chart at the book and he remembered it.

Try to imagine such a child at Kindergarten or at school. There are not so many teachers there, pleased when student knows things better and corrects what his teacher said. Gifted children need good educational programs and teachers from the very beginning of their schooling if we want to have enough adepts for the science at high schools.

Problems GC have at school I mentioned two years ago at my paper " Future Scientists and the School Attendance" (NATO Workshop in 2002). There was a 16 years old student of prestigious "Gymnasium" (grammar school, for students 11 to 19 years old), who was dissatisfied with the way his school cares for gifted students. Now, two years after, this student who was successful at several types of Olympiads, attended DSA, winned at the national level of competition in biology and was invited to meeting of winners to Australia, had the same problems with his school as two years ago. He asked his school for the Individual Study Plan and it was not easy to receive it. As he told us at Club for parents of GC meeting, what he really needed from the school was freedom in the way of his study, namely at the biology, where he starts to be expert and his knowledge is much more higher than that of his classmates or even of teachers in some topics.

Similarly a girl from another prestigious gymnasium (outside of Prague), excellent student interested in mathematics and highly successful in it asked for ISP but nobody wanted to speak with her. She wanted more demanding teaching of math, but school authorities felt hurt that she was not satisfied enough with the fact that she was excellent student of prestigious school. After two years of vain effort to make the school to a dialog the student decided to leave and now she attendes another school.

From the point of view of Ministry of Education or other institutions there were no problems. Both students were highly successful, with excellent school marks, without behavioral problems and moreover successful in competitions.

Many students have problems with overdirective treatment at schools. This is a great problem namely for highly gifted students,with their strong desire and ability to be independent. Schools and even society itself, even though they consider themselves to be open-minded, often follow a path that prevents development of self-confidence and competence. Authoritative leadership, or in best case "benign dictatorship", that does not allow for individual decision-making unless it conforms to the established authorities, inhibits development of one´s initiative and responsibility. As the result, it leads to apathy and to escapism (emigration, alcohol/drugs abuse, suicides, illnesses), or – in the case of stronger individuals – to anger and rebellion. Many individuals, perhaps most, try to blend in just to avoid conflict. Unenlightened teachers and psychologists (according to Kohlberg) actually consider this to be a demonstration of social intelligence(!).

Several years ago homeschooling became one of the ways how to educate children in the Czech Republic. There are not many families engaged in it. But their experience and children´s school achievement is very good. Children are more motivated to learn and because they cannot be "hidden" in the group of classmates parent as a teacher knows whether the child is ready to continue in learning new things without "gaps" or not. Homeschooling was partly used by families of highly gifted children I have met in recent years.

Two years ago 16 years old Tomasz Macura and his father participated in the NATO Workshop in Visegrad. Tomasz finished his university studies at that time. Now he continues in PhD. study at the Trinity College, Cambridge. Macura´s family came to USA when Tomasz was 6 years old. When he was 11, Tomasz graduated and was accepted to university, without attending high school. There are many myths on GC. Very popular one is that accelerated child cannot be "normal" in another ways, for instance social adjustment and physical health. Despite it Tomasz is well socially adjusted, likes sports, and is very friendly. For more information search "Tomasz Macura".

Last summer Czech-American family asked me for advice in their 7 years old son´s education. They visited Center for the GC in Denver and received very long and detailed report. Nick´s IQ was 190 and specialists suggested 25 recommendations what to do to realize his potential fully. One of them was to change the school and to go to one of the two best schools for the GC in the USA. By all means the family had to move. Boy´s mother came from Prague. That is why they hesitated if not to go there.

Preparation of ISP (Individual Study Plan) for that child was challenge and very nice task for me. I also introduced Nick´s parents to Macura´s family by means of e-mail. Realizing Nick´s education according recommendations from the Center in Denver was not fully in accordance with system of schooling in the CR. At the end the family decided to stay in the USA because of father´s job.

I have met several extremely gifted children during my practice. Being educated in the Czech system they attended probably more demanding mainstreem schools than the boys mentioned above. On the other side it was not possible to prepare and realize similarly individualized plans for them. Their possibilities were and still are much more limited. That is why such children are often boring at their schools sometimes despite the fact they were accellerated or went to another (better) school.

International exchange of information on GC education is very useful. We can avoid doing mistakes which prevent GC from developing their potential fully.There are also many good ideas for inspiration and cooperation. Very nice projects supporting gifted high school students interested in natural science is "Arachne" - summer camp organized by PhD. students

Ondrej Koukol and Jan Mourek from the Faculty of Natural Sciences of the Charles University. Authors informed on it at the 9th ECHA conference in Pamplona.

At the previous NATO Workshop I informed on a very good way how to prepare children for the scientific thinking and to familiarize them with research work. The Natural School in Prague is in advance, realizes ideas we are only speaking about and uses effective methods of education we know only from very good out of school programs. Inner motivation is needed if we really want to improve results of our work. Not only formal "looking like".

Above I was speaking on gifted and highly motivated students of prestigious gymnasiums (grammar schools). The schools were not interested in these students and did not support developing their potential and motivation enough or at all. I believe there are many students who experience similar problems. It would be useful to help them. For instance to involve highly able and motivated students to a nettwork which enables to help them if needed. Individual Study Plan developed in cooperation with University or other institutions educating students for scientific work. Some courses for school management and school inspection would be useful to inform them why and how to support gifted students. To develop some form of pressure on the school if needed (for instance in the case the school does not want to give a student time to participate at workshop, training course etc.). Educating the students not only in science but also in managing problems and cooperation with other students, teachers, organizations, institutions. And where to look for a help if he/she is not able to manage problems.

P. Csermely et al. (Eds.)
IOS Press, 2005


# Work with Gifted Students for Physics in the Czech Republic

**Zdenek KLUIBER**
*Ekogymnazium, Prague, Czech Republic*

**Abstract.** The contribution summarizes the experiences of teaching gifted students in the Czech Republic from the special point of view of physics education and the involvement of student in research practice. It lists ten demands (commandments) for the education of gifted children, gives methods for their selection and involvement in research with special reference to physics competition and teaching.

## Introduction

Along the increasing interest of students in the study of law, medicine, economy and philosophy, at present secondary schools in our country (and in abroad too) we can see a certain decrease of student's interest in physics, science and technical fields. However, at specialized gymnasiums it is possible to find a number of students who, thanks to their interest in physics (and mathematics), achieve very good study results as approved by their success in international competitions. These students are generally bent to become scientists in physics, let us say, physics teachers. The important consideration for the fulfillment of this resolution is that the students, during their university studies, can get a possibility to acquire not only physics knowledge at the appropriate level corresponding the secondary school curriculum, but foremost a possibility to display their creative activity simulating the work of scientist, let us say, a teacher.

For the technical and economical future of the Czech Republic we need physics, chemistry, molecular biology, informatics and engineering education as well. The representatives of the Foundation Chart 77, Czech Technical University, the Science Academy and Charles University published a joint pronouncement in which they apply to politicians, deputies, government establishments, scientist, university and secondary school teachers, representatives of banks and companies, entrepreneurs, representatives of the Nongovernmental Organizations and general public for help with searching and the development of talents for the Czech science and engineering, for help with enhancement of science and engineering, with supporting a viable initiative of the work with the young, with supporting the most varied forms of the intensive penetration of the latest scientific knowledge into searching, education and development of talents for science, into the life of the society.

Nowadays secondary schools providing general secondary education arc emphasized in the Czech Republic. There is an effort to increase number of secondary school graduates or university students studying at least bachelor‘s degree program.

The enactment of a new Education Law in the Czech republic means that not only national school-leaving exams will be put into effect but also the entrance examination for students of eight-year secondary schools providing general secondary education who want to continue their studies on the higher level. The new Education Law enables secondary school headmasters emphasize school subjects or curriculums that profile that particular school.

A peremptory characteristic, besides the corresponding professional level executing the work with gifted students, is time. Work with gifted secondary school students means not to give the time, but to try to envisage that time on the level of the cooperation: teacher X student-s. Both sides have to know what is the sense of their co-operation, mutual confidence has to exist between them.

The grasp and development of talents at secondary schools belong to the most important aims pertinent to the preparation of future scientists and experts. Just the decision of a secondary school student regarding his/her future, let us say, working orientation has to stem from the corresponding identification of the qualified work in the given field – according to the results of the research coordinated by the Centre for the research work in physics education of the Academy of Science CR1.

The assurance of this demand in the relationship to the grasp and development of gifted students especially insists on the fulfillment of the following targets:

1. students should be enabled their own creative work;
2. students should acquire the experience in the solution of simple research tasks within incorporating into the team of investigators;
3. within a teaching process it is necessary to involve an independent study of professional and scientific - popular literature;
4. students have to acquire the experience from the study of corresponding foreign literature;
5. students have to be enabled an appropriate contact with scientists to acquire actual knowledge about the results of the Czech physics research;
6. for gifted students it is needed to organize optional seminars of high quality;
7. to enable students to learn scientific discussion, forms and methods of scientific work;
8. to create conditions for the meeting of gifted students from different extracurricular activities in the given field not only in our republic, but also abroad;
9. at the corresponding rate and appropriate level, physics curriculum has to intercept the modern physics knowledge as an important motivation;
10. work with guide students involves the assurance of the organizing and material respect of the highly effected preparation of the gifted secondary school students.

## 1. Search for Gifted Students

A significant target of a gymnasium is to intensify individual work with talents. The fulfillment of this target supposes the individualization of education together with take in the abilities of every individual into account. A student's activity should come out of his/her inner needs with the aim to fill his/her self-realization.

On the other side, there is an indispensable circumstance, in what atmosphere it is possible to lead students to their success. The school atmosphere - it is our experience - plays a very important role in this way because school is able to ensure good conditions for these students' activities, in ail respects.

A peremptory assumption for the first stage of the work with students is the demanding physics teaching, work with the whole class. All work of a physics teacher should mainly come out of a good physical background, of the good knowledge of pedagogy and psychology. Only on basic of student' own activity a teacher can get to know the characteristics of a gifted student on physics.

The development of creative abilities (cyclic principle) is made in physics by the sequence: initial facts - model - hypothesis - logical results - experimental checking of the results. This method calls for more time than a mere announcement of the theoretical conception of a certain physics knowledge. It is possible to apply that if you want to develop students during physics teaching. The headmaster's business is to make the conditions of the cooperation between students and their teachers, to evaluate it and assert creative activities within the whole school.

In the period of a certain depletion of natural resources, the only source of the economical growth is highly creative work in the fields of research, development, technology and its application in ail spheres of the life of society. The needs of the development of the interest of the coming generation in ail new and progressive is connected with the system precaution creation which helps at the low age as possible:

- to reveal potential talent;
- to develop it by modern interactive forms in the field of both formal and informal education;
- to make a natural interest in a further study in scientific and technical fields;
- to make the interest of scientific, technical and business spheres in the preparation of talents for creative working places;
- to make the social prestige of the work with children and the youth in this field.

Above mentioned tasks result from the Strategy of the Development of Human Sources for the Czech Republic, adopted by the government on 3 March, 2003.

## 2. Activities in Physics

1. For the development of gifted students at gymnasium it is necessary to be in a face- to-face contact with physicists working in various physics fields and with the workers who immediately apply the results of the physics research at their work. One of the possibilities for the fulfillment of this demand is to organize seminars with regular meetings and lectures. The peremptory target of this seminar is the assurance of an actual bit of information and potential consultants for the student's work in competitions.

2. The educational program of the Foundation Chart 77 „ASTRA 2000" was established for secondary school students. It is focused on searching and supporting gifted students in physics, chemistry, biology, ecology and informatics. The project is oriented to the future and is joint with the aspiration to share the systematic education of future experts in given fields.

3. Every independent work of the students of the gymnasium is aimed in physics: in appropriate way to present records. One of the possibilities how to acquaint the public with the results of students' works - at school and invited guests - is the students' conference in the physics field. Its organizing is a significant action in favor of the total conception of the preparation of gifted students in physics at school, especially if it fulfils an educational aspect too /besides a presentation character! [1-4].

4. The International Association for Youth, Science and Technology (AMAVET) organized already 10th national exhibition of scientific and technological activities of students. This national exhibition of' student activity is sponsored by Czech power company EZ and is under the auspices of President of the ACSR. The exhibition has very positive impact on communication abilities of students, which are particularly useful for future representation on similar exhibition abroad.

The AMAVBT is the lasting participant of Expo Science International (ESI), which is organized by MILSET. The Czech students take benefit results of their work, of professional activities, and of improvement of their knowledge of English language.

During secondary school studies - in physics teaching school, in physics activities, in the support of the participation in physics competitions - a gifted student in physics is systematically prepared to university studies. The teacher takes a distinct responsibility for the development of the student's individuality. Obviously the support and collaboration of the creative teacher with the students is necessary. The students' education often could not take off without it.

To put together a publication requires experts in foreign languages and good knowledge of professional terminology. This fact supports some additional preparation for university studies.

Creativity is the basis of the teacher's profession. Only a creative teacher can develop the creativity of the students.

Teaching that develops the creative thinking of students changes their motivation. The principal reward is not a grade or mark but the resolution of the problem or the fulfillment of the school assignment. Such teaching improves the aims of study to the better and more qualified mastery of common thought methods. Students then know better how to ask and how to argue meaningfully.

Every successful creative teacher, who is not afraid of risking and who is capable of dealing with problems independently, imaginatively and with a sense of humor, should be encouraged by his headmaster (principal).

## 3. Physics Competition

Secondary school students in the Czech Republic can participate in four difficult physics competitions within their extra-curricular work which, on the instant, links to physics teaching. They are: Physics Olympiad, Secondary School Special Activity in the field of Physics, Young Physicists' Tournament and First Step to the Nobel Prize in Physics. The participation in one arbitrary competition mentioned above is, according to a specific individual focus of every student, a good preparation for university studies of physics. Naturally very demanding is just the final preparation of multimember team of potential representatives from which presentation team will originate in the end. The work of research lies not only in bringing new pieces of knowledge but first of ail in the ability to present them in an understandable form to scientific community, possibly to listeners or readers. The work of research teams is considerably emphasized nowadays.

There exist several international competitions in which, beside excellent written results of task processing or possibly project, substantial attention is paid to presentation. These competitions take place in English language. It is thus possible to claim, that very good English, correct English terminology ability to discuss and argue are demanded from students. Not only correct English terminology is concerned, but also the used specialized phrase i.e. words typical for professional argumentation in a given branch.

## 4. Project ASTRA 2000

If we look around on our quickly changing society we see that the measure of people's value of success has basically changed. The profession of scientists, engineers or technicians is not as prestigious as the profession of lawyers or economists. Before the World War 11, the former Czechoslovakia belonged to the ten most developed countries in the world. We cannot only live from the tradition although we can build on it, if we are able to develop it in creative manner. Decreased interest in natural and technical sciences is not only a Czech problem, but it is a problem of whole world. In developed countries, this has been solved for several years by investing large financial resources in the education of technical intelligence and for the support of scientific talents not only by the state but also by sponsors. In science development, market is very inefficient regulating mechanism, whose delayed feedback may even lead to complete destruction of science [5].

The inspiration of this project comes from the experience in USA, Israel and Sweden. These countries realized the negative impact of the small interest of young people in natural and technical sciences. They decided not to wait till market mechanism induce changes the interest among the young generation from the academic branches to technical and natural sciences. A similar alarming situation also exists in our country.

The aim of the project? The aim is to increase the interest in education in the field of natural and technical sciences among young people, the popularization of science and technology, the organization of education, the support of talented students and beginning scientists. The collaboration among universities and secondary schools, the Academy of Sciences and industrial companies to organize training activities could enhance the interest of young people about natural and technical sciences and could support talented students and could help in the systematic education of future experts.

## 5. School Physics Teaching

An important task of a high school is the intensity individual work with talented students. The fulfillment of such a task includes individual education with respect to the abilities of each student. The student's activity should develop from his or her own needs with a goal to fulfill one's own se1frea1isation [6].

On the other hand, the background in which it is possible to lead students to successful results cannot be neglected. No doubt, school background, which assures suitable conditions for such an activity, is vital to this process. Only on the basic of the particular work of the student, is the teacher able to determine features of talented physics students.

Correct solution to written examination, exact analysis of physical phenomena in oral examinations, reports on interesting topics, experimental skills, proper interpretation of obtained results and tokens of physical thinking, ail enable us to identify talented physics students. On the basic of this experience, the teacher can evaluate what is the best competition or direction of individual preparation for each student. But the teacher must talk to such students! He must know him or her personally from professional and moral points of view, from the point of view of his or her expected study specification, as well as from expected study specification, as well as from expected occupational orientation.

**References:**

[1] Kluiber, Z.: The International Conference of Young Scientists. Physics Competition, Vol 2, No 2, 2000, p. 39-40.

[2] Kluiber, Z.: Secondary Schools Special Activity in the Field of Physics. Physics Competition, Vol 4, No 1, 2002

[3] Kluiber, Z. et.al.: The Development of Talents in Physics. Prometheus, The Union of Czech Mathematicians and Physicists', Prague 1995, ll2p.

[4] Kluiber, Z. et al.: 10th International Young Physicists' Tournament. MAFY, Hradec Králové 1998, 92p.

[5] Rosenkranz, J.: Project ASTRA 2000. In: 10th International Young Physicists Tournament. IBM MMT, Prague 1998, 46 p.

[6] Kluiber, Z.: Students Preparation in Physics in the Czech Republic. In: 10th International Young Physicists' Tournament. 1DM MMT, Prague 1998, 46 p.

P. Csermely et al. (Eds.)
IOS Press, 2005


# Parents' Concept about their Gifted Children

**Maria HERSKOVITS**
*Budapest Institute for Education,*
*Vas u. 8. H-1088 Budapest, Hungary*
*hersm@orange.mtapi.hu*

**Abstract.** At the Centre for Gifted in the Municipal Pedagogical Institute of Budapest we began a follow-up study with nearly 500 children, who took part in counseling. Beside the basic sociological data (age, gender, etc.) this paper presents the parents' concept about their children in two ways. Regarding the checklists they regard their children in the first line sensitive, friendly, cheerful and helpful, and their relationship as a great challenge, but satisfactory. We used an "Early Signs of Giftedness" checklist, and , we subjected the data from the E.S.G.C. to factor analysis in order to find more basic lines of giftedness. As the result of the factor extraction there emerged six main directions: Verbality, Spatio-Visual Creative Performance, Logic, socialization, Efficiency and Social Penetrance. Five factors can be interpreted as a talent profile of young individuals, it gives an opportunity to compare profiles to each other and to use the profile as a diagnostic-prognostic tool in the future. One of the factors (socialization) may call attention on the possible adaptation problems

The Centre for Gifted in the Municipal Pedagogical Institute of Budapest is available for those children, whose abilities are obviously prove to be above the average, and their parents or teachers expect help in their development and related problems. The Centre provides enrichment programs and summer camps as well [1], but its basic activity is pedagogical and psychological counseling. The parents' problems vary from choosing the proper school through asking for help for the child's optimal development to serious educational, psychological problems. After a few years of counseling we became interested in collecting information on the development of these children, and estimate the impact of their family and school on it.

With the help of the OTKA Foundation T29273, we began a follow-up study with nearly 500 children, who took part in counseling (interviews, testing) and/or attended our enrichment programs between 1994 and 1998. Beside the systematic analyses of their earlier data we make a structured interview with the 12-22 years old former clients, and they are tested with the Wechsler Adult Intelligence Scale (WAIS), the Big Five Personality Assessment, Level of Aspiration Test, Guilford Creativity Test. The parents are asked to complete a questionnaire.

At first, we analyzed part of the questionnaires consisting of 35 questions (1.), and the checklists consisting of 37 items (2.) completed by all parents at the first interview. The device was prepared for the purposes of the newly established Centre at 1994, based on the works Renzulli, Rimm, Landau, and our earlier experiences [2].

## 1. The results of the questionnaires

First we analyzed 200 questionnaires. Besides gathering the necessary sociological data, the questions are organized around the development and education of the child, the characteristics of the child through the parents' eyes, and the parents' attitudes to them. The questions are partly open-ended, partly checklist type. The first sign of giftedness mentioned by the parents was experienced mostly between the age 2-4 years, and the event was usually in connection with an amazing question, the performance of memory or logical thinking. Most parents were rather ambivalent with these phenomena, and especially ambivalent with the increased level of activity, intrinsic interests, and an early striving for autonomy, which many of them experienced. They also felt an increased responsibility for the proper development of the child.

The majority of the children (more than half of them) were 5-7 years old - the parents came around the date of the schooling. The proportion of gender is a 3 to 1, three times as many boy as girls. (Table 1-2)

**Table 1.** Distribution of genders.

| | |
|---|---|
| Boys | 153 |
| Girls | 47 |

**Table 2.** Distribution of ages.

| Years | Number |
|---|---|
| 3 | 2 |
| 4 | 12 |
| 5 | 28 |
| 6 | 47 |
| 7 | 31 |
| 8 | 15 |
| 9 | 12 |
| 10 | 18 |
| 11 | 9 |
| 12 | 10 |
| 13 | 5 |
| 14 | 5 |
| 15 | 3 |
| 16 | 1 |
| 17 | 2 |

This proportion is similar to the experiences of the educational services: the parents and schools have much more problems with boys than with girls. That is why we cannot differentiate, whether this general fact's impact is decisive, or the parents are more ready to find gifted their suns than their daughters, or really, there are more gifted boys than girls.

Anyway, our experience is that more boys than girls are sent to enrichment programs without asking for counseling too. The majority of the parents are highly educated (Table 3), although we have relatively much parents who attended "only" colleges providing Bachelor's Degree. This and the high number of parents without higher education show that a lot of families who are not part of the so-called "elite" ask for our help. They are not familiar with the means and ways of fostering talent, but appreciate very high intellectual excellence. Compared to the experiences of other educational services, fathers were present relatively much. The interest and participation of the fathers in the development and achievement of the child may have an important impact.

**Table 3.** Schooling of the parents.

| | Father | Mother |
|---|---|---|
| 8 years or less (elementary school) | 8 | 5 |
| Trade | 19 | 9 |
| Secondary school | 40 | 62 |
| Secondary+course | 6 | 5 |
| High school | 38 | 51 |
| University | 86 | 67 |

The most frequently checked item on the checklist (Table 4) was „sensitivity", which in itself is ambiguous, - it may mean the open-minded attitude, but may considered as a synonym of being easily hurt - but the accompanying items: „friendly, helpful, cheerful, initiating, reliable" refer to that the parents consider their children basically positive. The high frequency of the items „stubborn" and „tireless" rather shows possible problems with the parents. However, nearly half of the parents told that there are no difficulties with the child.

**Table 4.** Characteristics of the children.

| | |
|---|---|
| Sensitive | 162 |
| Friendly | 149 |
| Cheerful | 142 |
| Helpful | 135 |
| Stubborn | 109 |
| Reliable | 100 |
| Initiator | 100 |
| Tireless | 95 |
| Adaptive | 86 |
| Neat, Orderly | 44 |
| Reserved | 41 |
| Lazy | 31 |
| Selfish | 25 |

The parents found mostly enjoyable and satisfying their relationships with their highly able children (Table 5.). It seems, the are able to perceive the special challenge, and they can learn from their children. Although, the low proportion of the items referring on problems is a bit different than our experiences based on the interviews.

**Table 5.** How do the parents consider the relationship with the children?

| Well-balanced | 101 |
|---|---|
| Great variety | 101 |
| I can learn from | 65 |
| Amusing | 63 |
| Great challenge | 49 |
| Difficult task | 45 |
| Enjoyable adventure | 43 |
| Tiring | 36 |
| Fight | 16 |
| Tensed | 9 |

## 2. The results of the Early Signs of Giftedness Checklist

A 37-item checklist was used to survey the possible signs of early giftedness. (Table 6). The E.S.G.C. is built upon the behavioral, intellectual and personality qualities of the children who, through elaborate testing and interviewing, proved to be gifted or able above average. Not all items are concerned with giftedness, there are a few, which are important for having information about the possible problems, that is why we do not use a total score. 420 children were surveyed for this part of the project, 114 girls and 306 boys. The average age of the children was 7,5 years with a standard deviation of 2,6 years. One-sample T-test was used to verify the below data, the significance level is marked + for ($p<0,1$), * for ($p<0,05$) and ** for ($p<0,01$).

The parents' answers showed that:

a) The gifted children are **high above average** in:
   i. Richness of vocabulary**
   ii. Speed of thinking**
   iii. Making correct observations**
   iv. The frequency of asking questions**
   v. Memory capacity**
   vi. Richness of imagination**
   vii. Curiosity**

b) They are **below average** in:
   i. Clumsiness**
   ii. Conventionality**
   iii. Shyness**
   iv. Difficulty in decision making **

c) They are **not different** from the average in:
   i. Liking of school / kindergarten
   ii. Self-esteem
   iii. Shifting everyday routine tasks onto others

**Table 6.** Early Signs of Giftedness checklist (E.S.G.C.).

| Please describe your child circling the appropriate number for each statement<br>3 – MUCH TYPICAL OF HIM/HER<br>2 –NOT MORE TYPICAL THAN OF OTHER CHILDREN<br>0 – NOT TYPICAL AT ALL | | | |
|---|---|---|---|
| 1) HAS A RICH VOCABULARY | 1 | 2 | 3 |
| 2) THINKS FAST | 1 | 2 | 3 |
| 3) HIS/HER MOTIONS ARE RATHER CLUMSY | 1 | 2 | 3 |
| 4) WANTS TO KNOW HOW THINGS WORK | 1 | 2 | 3 |
| 5) HAD AN EARLY INTEREST IN THE LETTERS OF THE ALPHABET | 1 | 2 | 3 |
| 6) KNOWS HIS/HER ENVIRONMENT VERY WELL | 1 | 2 | 3 |
| 7) LIKES TO TAKE THINGS TO PIECES | 1 | 2 | 3 |
| 8) DRAWS APPROPRIATE CONCLUSIONS QUICKLY | 1 | 2 | 3 |
| 9) HAD AN EARLY INTEREST IN NUMBERS | 1 | 2 | 3 |
| 10) MAKES CORRECT OBSERVATIONS | 1 | 2 | 3 |
| 11) ASKS MUCH | 1 | 2 | 3 |
| 12) HAS A REMARKABLY GOOD MEMORY | 1 | 2 | 3 |
| 13) HAS A VIVID IMAGINATION | 1 | 2 | 3 |
| 14) CURIOUS, RECEPTIVE TO NEW THINGS | 1 | 2 | 3 |
| 15) LIKES CONVENTIONAL SOLUTIONS | 1 | 2 | 3 |
| 16) HAS A GOOD SENSE OF HUMOUR | 1 | 2 | 3 |
| 17) TAKES RISKS, GOES INTO UNCERTAIN SITUATIONS | 1 | 2 | 3 |
| 18) HAS UNUSUAL, PERPLEXING QUESTIONS, IDEAS | 1 | 2 | 3 |
| 19) HAS A CRITICAL MIND | 1 | 2 | 3 |
| 20) TRIES TO SHIFT EVERYDAY ROUTINE TASKS ONTO OTHERS | 1 | 2 | 3 |
| 21) IMPULSIVE, ACTS BEFORE THINKING IT OVER | 1 | 2 | 3 |
| 22) ARRANGES THINGS INDEPENDENTLY | 1 | 2 | 3 |
| 23) SETS HIS/HER OBJECTIVES HIGH | 1 | 2 | 3 |
| 24) ATTEMPTING A TASK THAT INTERESTS HIM/HER, HE/SHE ALWAYS CARRIES IT THROUGH | 1 | 2 | 3 |
| 25) OVERSENSITIVE | 1 | 2 | 3 |
| 26) LIKES SCHOOL/KINDERGARTEN | 1 | 2 | 3 |
| 27) HAS A GOOD SCHOOL/KINDERGARTEN PERFORMANCE | 1 | 2 | 3 |
| 28) AGGRESSIVE, PICKS QUARRELS | 1 | 2 | 3 |
| 29) UNOBTRUSIVE, DIFFIDENT, SHY | 1 | 2 | 3 |
| 30) STUBBORN, OBSTINATE | 1 | 2 | 3 |
| 31) CONCERNS HIM/HERSELF WITH TYPICALLY ADULT PROBLEMS | 1 | 2 | 3 |
| 32) CONFIDENT, HAS A HIGH SELF-ESTEEM | 1 | 2 | 3 |
| 33) IT IS USUALLY HARD FOR HIM/HER TO COME TO DECISIONS | 1 | 2 | 3 |
| 34) ONLY SATISFIED WITH THE PERFECT | 1 | 2 | 3 |
| 35) TRUTH-SEEKER, HAS A SENSE OF JUSTICE | 1 | 2 | 3 |
| 36) DIFFICULT FOR HIM/HER TO TOLERATE FAILURE OR UNSUCCESS | 1 | 2 | 3 |
| 37) HIS/HER COMPANY TENDS TO FOLLOW HIM/HER | 1 | 2 | 3 |

There are also gender differences in parents' perception of gifted children, however these differences seem to be in accordance with gender differences of this age in general. Independent-samples T-test was used to verify the below results.

a) **boys** are described as:
   i. more clumsy**
   ii. more interested in how things work**
   iii. more likely to find their way in their environment well**
   iv. much more likely to enjoy taking things to pieces**
   v. slightly more interested in numbers*
   vi. slightly more imaginative*
   vii. more likely to pass their everyday tasks onto others*
   viii. more impulsive*
   ix. more likely to get into fights**
   x. slightly more stubborn +
   xi. slightly more interested in "grown-up matters" +

b) **girls** are described as:
   i. more independent**
   ii. more likely to love their school*
   iii. having better school performance**
   iv. having less difficulty making decisions**
   v. having less difficulty tolerating failure or unsuccess*
   vi. more likely to have their company follow them*

Age does not seem to have an effect on qualities surveyed with E.S.G.C. There was practically no correlation between age and the 37 items (the Pearson-r never exceeded 0.2), which makes us think that these characteristics are relatively independent of the passing of time in this period, therefore they may constitute **traits** of giftedness.

To test this hypothesis, we subjected the data from the E.S.G.C. to factor analysis in order to find more basic lines of giftedness. As the result of the factor extraction there emerged **six main directions** towards which the individual items of the checklist aggregated:

**I. VERBALITY, Cr-α =0.65**

Consists of items such as rich vocabulary (1), early interest in the alphabet (5) and in numbers (9), asking many (11) and unusual (18) questions – many of which are concerned with "grown-up matters" (31).

**II. SPATIO-VISUAL, CREATIVE PERFORMANCE Cr-α =0.6**

Consisting of items such as finding out how things work (4), being well informed and easy in immediate physical and social environment (6), taking things apart (7), being curious and receptive to new things (14).

**III. LOGIC Cr-α = 0.56**

Consisting of ease in drawing conclusions (8), having apt observations (10), being critical-minded (19) and sense of justice (35).

**IV. SOCIALISATION Cr-α =0.65**

Consisting of **reverse** items such as risk-taking (17), shifting tasks upon others (20), impulsivity (21), aggressiveness, and stubbornness (30).

**V. EFFICIENCY Cr-α =0.6**

Incorporating high objectives (23), persistence (24), perfectionism (34) and good school performance (27).

**VI. SOCIAL PENETRANCE Cr-α =0.6**

Made up from items such as liking of school (26), having self-confidence (32), having others follow (37), and being unobtrusive (29, reverse).

Concerning reliability, all six factors have been put through item-analysis which showed that all six factors have a Cronbach-alpha value considerably high. The factors themselves have a low correlation coefficient when correlated one to another, but the entire questionnaire has a 0.71 Cronbach-alpha value.

Some items, for example humor (16), memory (12), being quick-minded, failure-tolerance (36), conventionality (15, reverse) seemed not to fit in any of the six directions, some of them probably because they are too general {memory, speed, etc} or describing a rather specific area {humor, failure-tolerance, originality, clumsiness}.

Five factors can be interpreted as a talent profile of young individuals, it gives an opportunity to compare profiles to each other and to use the profile as a diagnostic-prognostic tool in the future. One of the factors (socialization) may call attention on the possible adaptation problems.

**References:**

[1] Herskovits, Maria: Developing Programs for Science-minded Children at the age of 7-12, in: Science Education (ed. Csermely, P. ad Lederman, L.) IOS Press, Amsterdam, 2003

[2] Herskovits, Maria: Family Influences at the Counselling Centre for Gifted, in: Gifted Education International, Vol. 4. No.2. p. 36-247, ABA, London, 2000.

*Science Education: Best Practices of Research Training for Students under 21*
*P. Csermely et al. (Eds.)*
*IOS Press, 2005*



# Research Training Program in Primary Schools of Hungary

**Péter JEAGER and József ZSOLNAI**
*University of Veszprém, Institute of Educational Research,*
*Jókai street 37,H- 8500 Pápa, Hungary*
*peter.jeager@freemail.hu*

**Abstract**. It is unjustified to separate science as knowledge creation by the elite from education as transmission of that knowledge to ordinary people. Instead, there is no basis of not providing 10-14 year old pupils educational opportunities to do research as an organic part of their schooling. Recognising this and considering other arguments for it Prof. Zsolnai and his team launched an action research project in 1998 in which talented pupils aged 10-14 were trained for research and did creative research. After 6 years into this project evidence collected indicates that it is a viable and highly valuable activity which develops many abilities and aspects of pupils' character. It is argued that research should be a part of everyone's schooling not only to develop their abilities but to meet the needs of a researcher-based society of the future.

## 1. Theoretical background

### a) Scientific research, knowledge and education

Scientific research (in the broader sense) is an *activity to create knowledge*. Is research an activity for scientists (working in the ivory tower) only? Why should we believe that scientific research (as a form of knowledge creation) is unsuitable and/or useless for anyone else? Why could not an ordinary person create knowledge himself for himself or for others? The methods of scientific research are often very simple. It is argued here that the image of scientific research as an activity suitable for the privileged few only is a belief taken for granted and it is totally unjustified.

At the dawn of mankind when language was not available for communication knowledge creation about the environment was almost entirely experiential and the resulting knowledge was *personal*. There was a direct link between individuals' knowledge and their environment via their experiences.

For better chance of survival knowledge needed to become transmittable, in other words, mobile in space and in time. The development of language as a mediating tool between people has made this possible. But it has resulted in something more: it has created an increasingly *impersonalised*, and publicly more and more available form of *explicit knowledge* which has gradually become a more and more prominent *mediating interface inserted between people and their material and social environment*. Public explicit knowledge creation has become a social role of the privileged few, and other people were given the opportunity to access and learn that formalised, explicit *ready-for-digestion* knowledge by education in a rather passive way. This has led to the development of *scientific research* and *education* (formal teaching and learning) as separate and distinct activities in their own right. Do these activities have to be so separate as they are now? If one accepts the positivist paradigm that science discovers truth and justifies it then making publicly available formal explicit knowledge a predominant source which people can use to gain and absorb information from is a sensible model to be followed, as it makes people's knowledge closer to that truth and it also saves a lot of time to find that truth.

What is the problem then with this, nowadays predominant way of learning as accumulating pre-digested and well presented information? It is not argued at all that mediation as such is wrong but arguably there is a danger of *intellectual poverty* if not only tools that mediate knowledge (such as language as it is considered in Vygotskyan theory [1]) are socially pre-constructed but also people's personal knowledge itself is largely pre-constructed into a digestible, clean form and thereby almost entirely mediated by some knowledgeable experts. Indeed, if we want changes we need to create appropriate mediating tools for learners that enable them to interact with their material and social environment in a manner akin to professional researchers.

Empowering people to do scientific research offers them:

- a new potential sources of information to create personal knowledge from, namely, the material and social environment, without losing benefits of systematic scientific research (that has proven to be of high practical value);
- a new form of behaviour and thinking that approaches both the material and social environment and ready-made public knowledge critically and inquisitively.

Therefore this resurrected direct, interactive, cognitively guided contact between people and the world around them is not a return to the past but it is an enrichment of the present with an additional form of gaining information and learning by personal knowledge creation. This personal knowledge does not necessarily mean something novel for the society, but knowledge novel for the person who creates it or for the social group which he is part of.

**b) Factors promoting introduction of research activities in schools**

There are some *philosophical movements* (which indirectly form public knowledge) that question the validity of scientifically created knowledge, and paradoxically they help introduction of research in schools because they suggest that explicit public knowledge which is largely a product of the sciences should not be taken for granted. These are:

- *scepticism* and *empiriocriticism* that question empirical methods of science to gain true knowledge (e.g. Hume, Mach);
- *critical rationalism* and *neopositivism* that question logical methods of reasoning widely accepted in science and thereby the "absolute truth" scientific research creates (e. g. Popper, Carnap, Feyerabend);

- *postmodern* that question rationality itself and the existence of truth the sciences aim to find.

In the area of *education* all movements that attack the notion of learning as absorbing ready-made knowledge are potential allies in introducing scientific research into school experiences, such as

- *educational constructivism*: effective and meaningful learning is a process in which pupils construct their own knowledge themselves instead of merely absorbing knowledge provided for them;
- *education for democracy* (e. g. Dewey) and *for critical reflective thinking*: members of a society should be empowered to become independent and effective decision-makers in a democracy, acknowledging the Kantian view that young children see the world from the subjective viewpoint and "the educational challenge is to help them understand that their own perspective is one among many and that truth, morality, etc. depend upon this understanding" [2];
- *problem-based learning and the process model of educational planning* [3]: problem-based instead of fact-based education, accordingly process-centred instead of content- or outcome-centred approach to education into which child research activities fit well;
- *child-centredness and individualism*: education should offer opportunities for all pupils to develop their own abilities in their own way, and this includes research activities if that is found to be appropriate in development;
- *gifted education*: closely related to child-centredness, aiming to provide educational opportunities for the gifted to build upon that giftedness (according to Prof. Zsolnai, everybody is gifted in something and everybody is disabled in something);
- *creative education* [4]: children should be given opportunities to do creative activities (and not only repetitive activities) at school to develop their creative abilities and attitudes (and research as knowledge creation is a creative activity);
- *experiential learning by reflection-in-action* [5] *or by implicit learning* [6]: effective learning that facilitates successful practice should be based on experience rather than on presented, pre-digested knowledge;
- Vygotskian *social constructivism*, contending that educational intervention can have a vast room to influence the quality of thinking children achieve [7]

In addition to the theoretical “push” towards research activities in education provided by these movements, there are also economical demands that “pull” education to that direction:

- *postfordism and the need for flexible, active workforce* [8]: for an increasing number of workers working is no longer a mechanical, rule-following process, instead, workers at any level of the hierarchy are expected to find information themselves by a flexible, inquisitive and interactive approach in order to solve problems;
- *lifelong learning* [9]: employees are increasingly expected not only to find ways to act autonomously in a situation but also to learn purposefully in order to be able to act autonomously; and research activities at school obviously educate pupils for learning purposefully and effectively.

### c) Factors hindering introduction of research activities in schools

There are some factors that act as barriers for introducing research activities in schools.

- *vulgarised approach to sciences in schools* [10]: natural and social sciences are unfortunately represented in schools in a distorted form in which only outcomes as ready-made pieces of unquestionable knowledge are covered, not its processes of knowledge-making, and open-ended enquiry is rare (but that is not a surprise given that teachers are totally inexperienced in research);
- *mass teacher training* [11]: as a consequence of (increased staff/student ratios), the image of the sciences at an undergraduate level of universities is devoid of its research aspect which would require personalised contact, and research-oriented training exists only at the postgraduate level;
- prevalence of *content-based* and *behavioural objectives-based* prescribed curricula in schools [12]: the open-endedness of research activities do not fit into such approaches of curriculum planning;
- *pressure for fair assessment of pupils and standardisation* in schools [4]: it is hard to make assessment of open-ended activities such as research defensible to prove that it is impartial;
- *Neo-Piagetian developmental psychologists* [13]: these experts contend that there is a direct connection between age and the appearance of formal thinking, and therefore there are age limits in successful acquisition of high level cognitive operations needed for research.

### d) Child researcher activities reported in the literature

One can classify child researcher experiments in the literature into two categories according to their major purpose: educational or non-educational. There are many examples of the latter group, in which social researchers involve children in research actively as co-researchers in order to use them to gain an insight into children's life or into the social environment in which targeted children live [14]. Although it is possible to derive some useful methodological principles (e.g. the recognised need for regular review sessions) from them, they are less relevant in the educational context.

What is the point of children doing such activities as a part of their education? The reasons can be classified into two categories: advantages from the child's point of view (progressivist reasons) and advantages from the society's point of view (instrumentalist reasons).

Two branches of progressivism can be distinguished: the nurturing one, aiming to provide adequate environment for the child's natural development, and the Vygotskian social interventionist one that tries to achieve the maximum amount of development possible with the child. Whichever branch one belongs to, research activities undoubtedly have developmental potential in all the following aspects:

- cognitive development [15,16];
- metacognitive development, self-reflection [15] and self-control [16,17];
- development of communicational abilities [15];
- social skills development [16,17];
- improved attitudes to learning [15,17];
- improved self-respect and independence [15,17];
- improved organisation and management skills [18];
- developed ethical awareness [18].

Instrumentalist reasons are no less convincing either, and include all the following:

- it helps identifying and educating talented researchers of the future [17] which may have a vast impact of utilising human resources of the country more effectively;
- it helps improving public understanding of the sciences and scientific research [15];
- it improves respect for creators and creation as a valuable activity [15];
- it contributes to actively involving children in the still predominantly adult-led society [18].

Of course the idea of child research activities in education is not new but that is not a surprise given the amount of potential theoretical and paradigmatic support to it mentioned above. Indeed, child research activity has become a well known educational method in primary schools over the last five decades [19]. It is an especially frequently used method in gifted education. Nevertheless, findings do not always indicate that pupils' learning by research is necessarily efficient if adequate ongoing support is not provided [20]. One possible explanation is that the cognitive load concomitant to research activities might be unmanageable by some pupils working on their own [19,20]. As a remedy to that problem structuring those research activities has been found to be useful [19] and it seems that pupils need many forms of guidance and intervention during their research process to make it successful [20]. It is less frequent, however, to provide preliminary formal research training for pupils, but there are some successful examples of that [21,18].

## 2) The Zsolnai approach to educating children-as-researchers: Pupils' Scientific Circles

Prof. Zsolnai is a widely known educational innovator of Hungary, whose educational theory could be crudely characterised as the "pedagogy of hope" because he believes that every pupil can be formed and should be formed. His recommended approach to fulfil that hope can be characterised by the expression "enforced progressivism" as it concentrates on forming the child as an individual by developing his/her of needs (need for acceptance, status, appraisal, self-expression), emotions, abilities, behaviour, self-control, willingness, character, attitudes, creativity, self-awareness and desire for justice and democracy by appropriately chosen demanding activities and individualised education [22]. Children doing research is a kind of creative activity that fits very well into his pedagogy, since it seems to have a potential to develop all these aspects of the child simultaneously, and it is an activity that can be easily individualised. In fact, he and his followers believe that creative activities (such as research) should be made available not only to high achievers and talented ones but to all the pupils [17].

Moreover, Prof. Zsolnai has been arguing for a faithful representation of not only outcomes but also processes of sciences in schools since the 1970s [22], which he believes to be made possible by adapting our already existing metaknowledge about the sciences to school context appropriately [15].

### a) Preliminary events

Because child research activities harmonise with Prof. Zsolnai's approach to education, it is no wonder that it occurred in one of the primary schools following his principles (in a village called Zalabér). In 1997/98 90% of 6-14 year old pupils of that school did creative projects (including research projects) of their own choice and exhibited its product (research report or art work) to others (children and teachers). The best ones were given the opportunity to compete for prizes by presenting their work for an audience (children, teachers and parents) on "creator days" [16,17].

Experiences with the "creator days" have led to the idea that a proper scientific research training (preceding scientific research executed by the more talented pupils) could be useful for them because it seemed likely to enhance the quality of their research processes, thereby enriching their learning. This idea seemed bold at the time and it was vulnerable to many kinds of criticism [15].

**b) Method and hypotheses**

There were too many variables to control for them in the practical experiment ahead. Therefore an *action research* approach [23] with one year long cycles was chosen. Experiences from a given year was used to prepare and execute the next years' phase of the experiment. Of course the experiment could be called entirely successful if all the following hypotheses (worded by the researchers involved) proved to be right:

1) child research training is possible to do successfully, in other words, talented children from age 10 upwards are capable of learning and applying methods of scientific research and presenting their results;
2) teachers are willing and capable of helping pupils to acquire knowledge and skills of scientific research and supervising them in their research;
3) child research training can become an organic, sustainable part of schooling.

**c) The first cycle of the action research project: the 1998/1999 school year**

As a Pilot Study 17 talented 10-14 year old pupils of the school in Zalabér were offered *research training* which involved:

- learning and applying a step-by-step algorithm of doing research: choosing a problem, specifying it, placing its topic in the taxonomy of sciences, collecting data about the problem, forming hypothesis, collecting and organising factual data, creating appropriate model out of data, judging the truth of hypothesis, offering ways to solve the problem;
- learning how to use the library to distract and reference information;
- learning how to use computers to create research report and presentation slides.

Pupils were given opportunities to choose their topic of research in 3 broad areas: ecological problems, economical problems, tragic events of mankind. They were required to submit a 15-30 pages long research report [15]. Pupils were allowed to miss 20 regular school lessons provided that they work according to the schedule. At the end of the term they made a 10 minute long presentation of their research projects for a small audience of professional researchers who asked them questions and gave them feedback [16].

**d) The second cycle: 1999/2000**

The Pilot Study was repeated (including its training phase) in this year using a larger sample. 10-14 year old pupils of schools of Zala County were offered the opportunity to participate in the talented education programme by then named as "*Pupils' Scientific Circles*" (PSC) the same way as its then already existing counterpart in secondary schooling [24], and 58 of them volunteered from 6 schools, 31 of which completed the project by submitting a research report and giving a 10 minute long presentation for an audience. Pupils were given oral and written feedback by the assessors who had read their reports and listened to their presentation. This time pupils were allowed to *choose any topic* from natural or social sciences or arts. 83% of pupils followed the taught algorithm of research in their research reports, and almost half of them could classify their topic in the taxonomy of sciences. [16]

**e) The third cycle: 2000/2001**

In this school year the number of talented pupils (aged 10-14) participating in the " Pupils' Scientific Circles" have grown to 68 from several counties of Hungary, 48 of which completed their projects. By the end of the third school year experiences collected were judged sufficient to make steps for a considerable expansion.

**f) The fourth cycle: 2001/2002**

In the 2001/02 school year the "Pupils' Scientific Circles" competition became countrywide with more than 150 volunteer pupils from 25 schools. To cope with their increased number 3 regional and a national final round were arranged. To improve the quality of pupils' research training (as new feature) 13 *school teachers were trained* on a short course on how to teach pupils research skills and how to supervise them, and a *handbook for the participating teachers* was also produced by the researchers for the same purpose. The best 16 pupils of the 38 in the final round were invited for a *viva-like examination* in front of a committee which consisted of *professional researchers* who had prepared for the presentation by reading pupils' research reports and after the presentation they disputed children's research methods and results and researcher *pupils had to defend their ideas, models, methods and conclusions*; and (according to the committee) 6 of them did astonishingly well in that. Another important experience was that *pupils of teachers trained for research training did remarkably better* than those whose supervisor teacher was untrained.

**g) The fifth cycle: 2002/2003**

*Another 11 teachers were trained* for teaching and supervising research in this school year, and more than 200 pupils participated in the competition. This time pupils were required to do *self-evaluation* in their research report as well as in their presentation. In addition, the *handbook for teachers* was *updated* according to the experiences and another *handbook* was produced *for researcher pupils*.

**h) The sixth cycle: 2003/2004**

More than 300 pupils participated, and accordingly the number of regions were increased to 5 instead of 3. Pupils were asked to *present a synopsis of their work in German or in English* both orally and in a written form (which they did well). Performance of all 44 pupils in the national final round were criticised orally after their presentations by professional researchers and PhD students. Both handbooks were updated.

**i) Handbook for pupils**

The handbook for pupils is now 43 pages long and it covers the following:

- characteristics of science and scientific research
- taxonomy of sciences;
- types of research;
- steps of doing a research (choosing a research topic and a supervisor teacher, gaining background knowledge, exploring the problem, forming hypotheses, empirical and theoretical data collection, comparing data with hypotheses, presenting research report orally and in a written form)
- research ethics.

**j) Handbook for teachers**

The handbook for teachers is now 89 pages long and it provides an introduction into:

- creatology, knowledge creation and the potential role of PSC in a knowledge-based society;
- the contradictions and potential conflicts between school culture and research culture that supervisor teachers need to cope with;
- the aspects supervision of pupils should aim to develop in pupils and the importance of values in the process;
- the route from school projects to research;
- the formal contract which the pupil, his/her parents, the supervising teacher and the head teacher have to sign before undertaking PSC work;
- the possible forms of individual and group work with pupils in PSC;
- the ways the supervisor can increase the effectiveness of PSC work (self-criticism, increasing own competence, acquiring and applying action research methodology, identifying himself/herself with the ideal of a knowledge-based society);
- ways PSC work should be documented by the pupil (diaries and research report);
- ways pupils can improve their communication skills;
- requirements and important aspects of assessment of oral and written research presentations;
- the way rounds of PSC competition are arranged;
- ethical problems supervisor teachers may face and their pedagogical connotations (e.g. influencing pupils in the type and topic of research, the amounts and form of help they may be given).

### k) Results of the action research project so far

Results *strongly confirm all 3 hypotheses* of the action research project:

1) *Child research training is viable and useful* for high achievers of at least one subject from age 9 upwards. These pupils can acquire applicable scientific research skills by research training and they are capable to execute a scientifically sound piece of research and present it in a professional manner both orally and in a written form.
2) *Teachers can teach and supervise* child researchers effectively, especially *if they are trained* for that purposefully.
3) Child researchers' training and their activities may become an organic, sustainable part of schooling and it makes their learning more meaningful.

### l) Would you believe it? Some examples of pupils' works

In addition to the positive outcomes, one has to notice that the quality some pupils' work is far beyond what most people would expect from even a talented child between 9 and 14. Without exaggeration, some research reports would have a good chance to be published in scientific research journals. Here are some examples from last years' national final:

*Andrea Berekszászi (10 years old): Being and remaining Hungarian in Ukraine.* [An empirical research using questionnaires and interviews with Hungarians (children and adults) on both sides of the border, attempting to find reasons for the fact that many Hungarians and their descendants could retain their Hungarian identity after their homeland was annexed by other countries, and for the high importance of that identity to them. Findings point toward the essential role strong family relations play in keeping community folk traditions (connected to religious festivities) alive, thereby helping to retain identity and strengthen its importance.]

*Kitti Varga: (11 years old): Children in foster care.* [An interview-based research seeking to find answers to the following questions: 1. To what extent people differentiate between children growing up in foster care and others? 2. How much chance do abandoned children have to return to their real parents? Findings indicate a negative discrimination against these children by members of the society especially if they are native Gypsies and a mere one percent chance for abandoned children to return to their real parents partially due to their parents' poverty.]

*Tamás Horváth (12 years old): Examination of heart functioning in relation to physical effort.* [An empirical research of 143 pupils aged between 11 and 18 based on numerical data of their (systolic and diastolic) blood pressure and pulse in repose, immediately after physical effort, and after some rest following physical effort. Using statistical analysis (t-test) and diagrams data from different groups of the sample (boys and girls, those who do sports regularly and those who do not) are compared. Conclusions about differences detected are discussed and possible physiological explanations for those differences are offered.]

*Máté Aczél (13 years old): Bows of the migrant Hungarian tribes.* [A partly literature-based, partly experimental research questioning both stereotypical agreement in the literature that 10th century Hungarian tribes had far more effective bows than Western European nations (helping them to win battles) and experimental research methods applied by historians using replicas of ancient Hungarian bows on the basis of their authenticity. Conclusions include that Hungarian reflex bows were probably not better than other contemporary bows and therefore Hungarians' success are more attributable to an auspicious harmonisation of battle tactics, strategy, weapons and logistics; and that the low number and quality of findings from excavations make current replicas largely dependant on unreliable speculative methods applied by their makers.]

*János Orsós ($8^{th}$ year): Gypsies' housing conditions and consumer behaviour in Osztopán.* [An empirical (observation- and questionnaire-based) study of Gypsies' housing conditions, consumer behaviour, their expectations about their own and about their children's futures. It is put into a theoretical (social, economical, cultural) context citing depressing findings of earlier research (high Gypsy unemployment rate, incomes of families insufficient for bare living, lost Gypsy identities, diminishing folk customs) and contemplating about possible reasons for the growing gap and barrier between the Gypsy underclass and other layers of the society. Empirical findings include that their homes usually lack certain amenities; that they regularly buy only the most basic goods, but they also purchase fashionable but unnecessary goods such as mobile phones. The author (who is a Gypsy himself) points to education as the only chance for them to find jobs and get out of poverty.]

## 3) Conclusion: from knowledge-based society to researcher-based society

There are at least three lessons we can learn from the experiences of the action research described above that lead us beyond its original purposes:

- *children*, especially talented ones *might be capable of achieving a lot more than we assume about them*;
- children could be involved more in societies and *we should not consider them inferior to adults since there is no basis of that*, and it is time to throw away Piagetian dogmas of cognitive development;
- research does not necessarily have to be considered an esoteric activity, but *an activity which many people can master and do successfully if adequate training and guidance are provided for them.*

I believe that in a not too distant future all jobs will demand a researcher-like attitude to the environment and research-like activities from its workers. This is not a new idea; for example Stenhouse [3] advocated the idea of teachers as researchers almost 3 decades ago. Our experiment shows that educating for research can be started at a very early age and this might be one of the first cornet steps that contribute to the process that leads to a different society in which all people become active and creative researchers; and then "researcher-based society" will be a more adequate slogan than the current "knowledge-based society".

**References:**


[1] R. K. Parilla, Vygotskian Views on Language and Planning in Children, School Psychology International 16 (1995) 167-183.

[2] L. G. Splitter, Critical Thinking: What, Why, When and How, Educational Philosophy and Theory 23 (1991) 89-109.

[3] L. Stenhouse: An Introduction to Curriculum Research and Development. Heinemann Educational, London, 1975.

[4] NACCCE, All Our Futures: Creativity, Culture and Education, London, 1999.

[5] D. A. Schön, Educating the Reflective Practitioner, Jossey-Bass Publishers, San Francisco, 1987.

[6] A. S. Reber, Implicit Learning and Tacit Knowledge, Oxford University Press, New York, 1993.

[7] M. Shayer, Piaget and Vygotsky: A necessary marriage for effective educational intervention. In: L. Smith, J. Dockrell, P. Tomlison (eds.), Piaget, Vygotsky and Beyond, Routledge, London, 1997, pp. 36-59.

[8] D. Mankin, Toward a Post-Industrial Psychology: Emerging Perspectives on Technology, Work, Education and Leisure. John Willey & Sons, New York, 1978.

[9] SEED, National Dossier on Education and Training in Scotland, SEED, Edinburgh, 2003.

[10] J. Zsolnai, Az értékközvetítő és képességfejlesztő pedagógia, ÉKP központ - Holnap Kkt. - Tárogató Kiadó, Budapest, 1995.

[11] J. Furlong et al., Teacher Education in Transition, Open University Press, Buckingham, 2000.

[12] A. V. Kelly, The Curriculum: Theory and Practice, 3rd ed., Paul Chapman Publishing Ltd., London, 1989.

[13] R. Case, Neo-Piagetian Theories of Intellectual Development. In: H. Beilin and Pufall, P. B. (eds.), Piaget's Theory: Prospects and Possibilities. Lawrence Erlbaum Associates, Hillsdale, 1992, pp. 61-104.

[14] A. Jones, Involving Children and Young People as Researchers. In: Fraser, S. et al (eds.), Doing Research with Children and Young People, Sage Publications, London, 2004, pp. 113-130.

[15] J. Zsolnai, Kutatóvá nevelés már tízéves kortól, Magyar Tudomány, 49 (2004)

[16] A. Kiss, A TDK lehetőségei az általános iskolai tehetséggondozásban, Új Pedagógiai Szemle, 51 (2001) 88-98.

[17] Zs. Cseh Németh, Tudományos diákkör 10-14 éveseknek, Zalai Tanügy, 6 (2002) 15-16.

[18] M. Kellett, Empowering ten-year-olds as active researchers, Paper presented at the British Educational Research Association Annual Conference, Heriot-Watt University, Edinburgh, 11-13 September 2003

[19] D. P. Brien, The Teaching and Learning Processes Involved in Primary School Children's Research Projects. Dissertation Abstracts International, Section A: Humanities and Social Sciences 57 (1996), 0108.

[20] K. Bishop, The Research Processes of Gifted Students: A Case Study. Gifted Child Quarterly 41 (2000) 54-64.

[21] J. A. Rankin, The Effects of Summer Research Training Program on Science Process Skills, Self Efficacy towards Creative Productivity and Project Quality. Dissertation Abstracts International 54 485.

[22] J. Zsolnai, Pedagógiai jegyzetek műveltségről, tudományról. Köznevelés 27 (1971) 5-6.

[23] J. Elliott, Action Research for Educational Change, Open University Press, Milton Keynes, 1991.

[24] Z. Borsodi et al., Kutatási lehetőségek középiskolásoknak. Magyar Felsőoktatás 12 (2002) 12-13.

P. Csermely et al. (Eds.)
IOS Press, 2005


# Nationwide Mentorship Network for Talented High School Students in Hungary

**Katalin SULYOK, Anna SÓTÉR, Ferenc MINÁRIK, Máté SZALAY, László FAZEKAS and Tamás KORCSMÁROS**
*Hungarian Research Student Association*
*19-21. Ajtósi Dürer sor, H-1146 Budapest, Hungary*
*info@kutdiak.ph.hu*

**Abstract.** Here we present the social and scientific goals and achievements of the Hungarian Research Student Association and our experiences in the establishment and first few years of a nationwide organization. This student-based organization is unique as it has a huge mentorship network in Hungary and in the neighboring countries, and it is led by the students. It is among the largest movement of the world for talented high school students with its more than five thousand members.

## 1. Introduction, the history of the Association

In 1995 an unprecedented initiative has been started by Peter Csermely, professor of biochemistry at the Semmelweis University (Budapest, Hungary), which resulted in the official establishment of the Hungarian Research Student Association (HRSA, or Association) in 1999. The basic principle of the Association was to help the talented and motivated students in the ages between 14 and 20 (free of charge) to obtain first-hand experience of scientific research in Hungarian universities or research institutes.

The cooperation between talented students and the professors is helped by a book containing the list of the mentors. The 1996 list contained approximately 300 scientific laboratories, the number of which increased to almost 700 by 2004 (Figure 1.) from 42 cities of the country and also from Austria, Australia, Canada, Italy, Romania, Serbia and the United States.

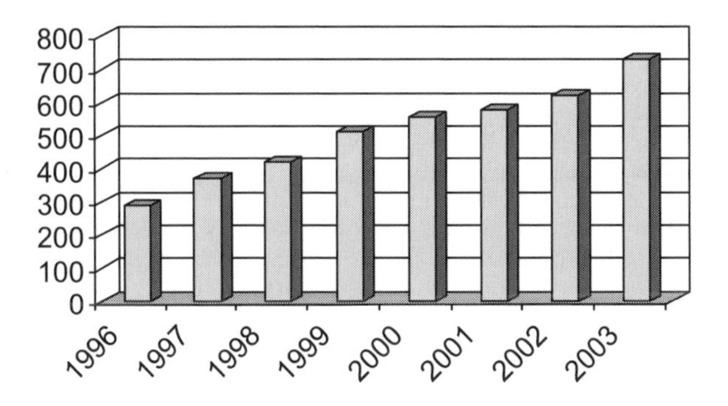


**Figure 1. The figure shows the change in the number of mentors in the period of 1996-2003.**

This list is originally published as a book and it is delivered to each member also via e-mail containing a wide range of keywords (approximately 3000, from abortion to x-ray micro-analysis) to help the students decide what their field of interest really is. Every high school headmaster gets this book annually in Hungary and in the surrounding countries as well, because we have several hundred students from neighboring countries (Slovakia, Ukraine, Romania, Serbia, Croatia and Slovenia).

Among the scientists who accept the students in their laboratories there are 118 members of the Hungarian Academy of Sciences including Nobel Laureate George OLÁH. The patrons of the program are Ferenc MÁDL, President of the Hungarian Republic, Bálint MAGYAR, Minister of education and Sylvester E. VIZI, president of the Hungarian Academy of Sciences.

In 1998 the Hungarian Research Student Foundation was established to manage finances such as the annual budget of the Hungarian Research Student Association. The movement is accountable and transparent: the decisions are made by the democratically elected student leadership and all affairs are publicized on the web-site. The main sources of support for the Foundation are funds, firms, the government and international organizations like NATO, UNESCO, the European Union and indirect donations via the Hungarian personal income tax system. From these sums we support establishing and functioning Research Student Clubs in the high schools. More than 200 such clubs were founded in Croatia, Hungary, Romania, Slovakia and Serbia. Members of these clubs cooperate in larger research projects and inform each other regularly of their progress.

In the fall of 2002 the Association opened an office in Budapest, where the coordinator of the movement works. The coordinator assists and helps in the work of the student presidency to organize the programs for the more than 5,000 student-members.

## 2. Membership of the HRSA

As a starting remark it needs to be stated that students do not take any administrative responsibility or duties when they apply for membership (besides the fact that they are responsible for their own progress: if they do not work properly, their enthusiasm or diligence is missing, they will not have any results); being a member is only an opportunity, which is free of charge.

In order to obtain membership in the Association students have to answer two very simple questions on our website (http://www.kutdiak.hu): "Why do you want to pursue research?" and "Why do you feel that you are better than other students?". We haven't got strict age limits in terms of the beginning age of the members but students may only apply for membership until their junior year at University or College. The youngest student we have enrolled so far was 11 years old, however, most of the students are between 17 and 19. As a matter of fact there is no selection of students. As a result of different motivations a simple self-selection is present in these groups. Maturity is a basic factor because the Association does not help the student to make the connection with the mentor: everyone builds the initial contact alone. Our mentors treat all students as equal partners. Clearly, not every interested student could easily get their own research project, as mentors have all rights to test them and their preparation for the project. There are labs, where the students have to pass oral or written exams of the books and research materials their mentors gave them.

In terms of educational backgrounds it is important to mention that the majority (more than the half) of the members lives in rural areas: a quarter of them come from small villages and another quarter from small towns. That is a good sign showing that the movement helps the spread of equal opportunities for socially underprivileged students as well as gives them the chance to change their social status.

It is worth mentioning that almost exactly 50% of the students are and were always girls, which shows a balanced interest for scientific research in both genders. However, the interest towards science and research also increases generally (Figure 2.).

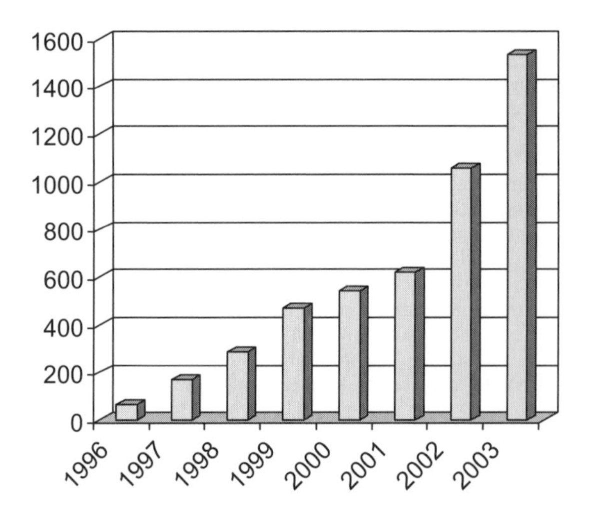


**Figure 2. The chart represents the increasing number of the individually registered student-members between 1996 and 2003 (an additional 2,500 to 3,000 members are working in research teams of the participating high schools).**

Almost 60% of the students work in life science laboratories, while 27% and 16% pursue their research in other natural sciences and social sciences/humanities, respectively. These rates, which have been more or less stable in the last 4 years, show the popularity of the life sciences among the youngest generation, especially in fields of environmental and medical research.

## 3. Annual meetings and conferences

A student researcher with a completed project has the opportunity every year to present the results at one of the seven regional conferences. After the evaluation of their research, the best third of the students is selected to present the projects on the National Conferences of Research Students (TUDOK), which has been organized four times since the year 2000. Outstanding figures of Hungarian science participate in the juries on these conferences evaluating the presentations and performances in different sessions: humanities, social and natural sciences are all to be found among these sessions.

A few of the best might take part in other international science-camps, and all the prize winners participate in the National Scientific Camp, first organized in the summer of 1997 in Káptalanfüred at the Lake Balaton. The best Hungarian scientists are invited here to talk about their approach and devotion to science just like a few respected writers, priests and successful business- or statesmen with whom the opportunity of discussion is open for the participants. The annual assembly of the HRSA also takes place here, where the president and the two vice-presidents are elected for the coming year.

We also have an essay competition once a year conducted by the movement's Human and Social Science Section. An Anthology was published from the award-winning essays in the winter 2004.

The 1st National Conference for Research Student Leaders was held in February 2004. It was an important step in the progress of the self-organization of the HRSA, because we have to extend and renew the leadership due to the very intensive turnover of high school student researchers (the average time a student spends in the Association as an active member is around two years). This is the only way to fulfill the basic feature of the Association: *"The students make it for themselves"*.

After the self-organizing student movement, a self-organizing science teacher movement is about to be born. There are approximately 600 high school science teachers who learn how to recognize and help talented students and form student-teacher research teams. Fortunately a major increase could be recognized in the number of science teachers helping to recruit students (Figure 3.).

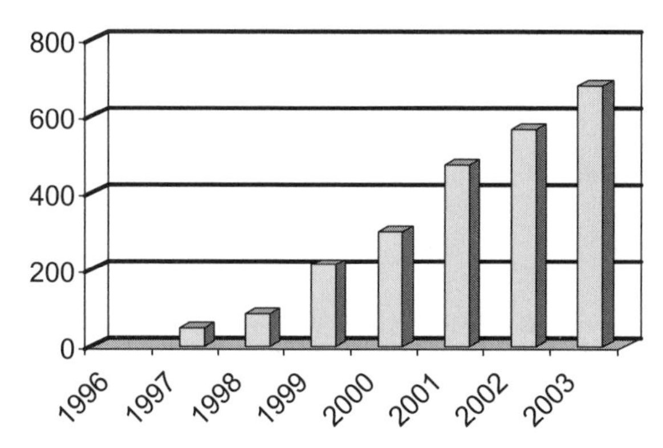


**Figure 3. This diagram shows the increase of the science teachers in the movement between 1996 and 2003.**

Some of the recent plans of the emerging Hungarian Research Teacher Association are to organize conferences for science teachers to exchange and discuss experiences and actual questions. This hopefully will help in the formation of a solid and strong Association in the near future.

## 4. Results and experiences

The idea of the Association immediately gained an overwhelmingly positive response from the Hungarian scientific community. We are now a medium-scale program, which proved its efficiency, sustainability and usefulness. The movement as a recognition to its useful activities in recruiting the future generation of scientists has received the 2003 Science Communication Award of the European Molecular Biology Organization (EMBO) and the highest award of the European Union for Science Communication, the 2004 Descartes Prize.

Our success is mostly due to our commitment to do only "real science" and not prepared pseudo-experiments: this gives real excitement and teaches the students to cope with real frustrations. They have to become a real member of a real scientific lab working on real research projects.

Other essential key-features are: the lack of bureaucracy, pre-set administrative goals and prescribed uniformity. The whole movement is highly personal and not administrative, and the style of handling with problems is playful trying to show that science is fun. We realized that this is the easiest way to attract young students to scientific research, which has a growing importance both in Europe and world-wide. As a matter of fact high school students are very susceptible and sensitive to new influences from outside their familiar environment. This is the time of self-discovery, when they explore their capabilities and limits and seek their place in a society, which is strange for them. They become more critical of established and strict social and familiar rules and other things that they once more or less just accepted.

For research students science gives a world without limitations and boundaries. Research is the best way out of the everyday routine, which is killing creativity in many schools. That is an additional reason to include students in science at this age. Due to their sensitivity, if students get to know research at this time, they are very likely to stay in this field later on. As an evidence for this we proudly note here that some of our first research students have already received their Ph.D., or have returned to their former schools as teachers helping to recruit the next generation of student researchers.

Thanks for them and our successes in the past few years we have gained quite a bit of experience how to attract young people to science and research. It is our goal in the future to spread this experience via the Network of Youth Excellence, which is about to be formed officially in this conference and by other means.

Last, but not least we have to underline that the Hungarian Research Students Association is loyal to its motto: *"A place, where you find yourself!"* We help young students to find themselves among the others similar to them, having the same fields of interest to establish a solid base for their future friendship and collaboration in research projects.

## 5. Future plans

We will keep organizing our conferences and meetings, making more efforts to recruit new students. We continue making our contacts stronger and more wide-spread abroad, and we would like to reach heretofore "uncovered" high schools in Hungary too. We try to make a network of activities to our students, helping the self-organization of Student Research Clubs, and various Sections of the Student Association. The help of the establishment of the Hungarian Research Teacher Organization is an ongoing project. If our heroic efforts to establish a stable and increasing financial funding will be successful, we are about to prepare to "export" our model to other interested countries in Europe and other parts of the world. As another long-term project we want to be a part of a strong alliance of talent support activities both in Hungary and world-wide.

*P. Csermely et al. (Eds.)*
*IOS Press, 2005*


# Biotech Program for High School Students in Hungary

**Ágnes TÁTRAI**
*Bio-Science Ltd.*
*Andor utca 47-49. 1119 Budapest Hungary*
*tatrai.agnes@bio-science.hu*

**Abstract**. The fast and new developments of genetics have generated a need for knowledge in the society. It is especially obvious among youngsters, the future generation. We have decided to fill this gap. A profit-oriented company, Bio-Science Ltd., Budapest teamed up with Semmelweis University, Budapest, and organized a summer camp for the high school students in genetics. The courses were extreme success suggensting that such a collaboration may serve for the benefit of all participants.

## Introduction

The 21st century is the century of genetics. It is a dynamically developing science influencing and determining our everyday lives. To draw attention and communicate the development of sciences including genetics is also the role of high schools. What is needed to accomplish this aim? Beyond infrastructure, the knowledge and capability of teachers to transfer the achievements of science are absolutely essential. Even in well-developed countries the source of funds is not proper enough to provide and introduce the newest technologies in secondary level education, though, it is within the scope of duties of secondary schools to improve talents with additional trainings and courses. This is even more important in biology and genetics. These sciences are developing so fast, introducing new ideas, methods and techniques that the public education – even if the funds would be enough - cannot compete with the speed of changes. What can a possible solution be?

Every country has its own traditions, sources and prospects according to which the possibilities of educating children can be exploited. In Hungary, the government is not able to expend enough support to specialized summer courses, therefore it is the role of profit-oriented companies to create and finance these kinds of traditions.

## 1. BioGen Summer Camp

Bio-Science Ltd. has decided to organize a summer camp for high school students in genetics. The basic idea was to transfer the most up-to-date general knowledge in molecular biology and genetics in the form of lectures as well as practice. Bio-Science Ltd. being a profit-oriented company with a main profile in biotechnology has recognized the need and offered support to organize the BioGen summer course with the help of Semmelweis University Medical School providing an academic environment.

We started a unique and exemplary summer program called BioGen Summer Camp in summer of 2003 and continued it in 2004. We organized two, one week long genetics courses in each year for secondary school students interested in biology who were intended to deepen their knowledge in the field of genetics replenishing their secondary school studies. The students, besides attending lectures, could analyse their own and some unknown DNA-samples with different laboratory methods within the confines of research programs and also became acquainted with high-tech instruments.

Bio-Science Ltd. undertook the following scope of duties in organizing and performing the summer camp course:

- All tasks regarding organization and performance (applications of students, mailing, organizing entrance exams, setting up labs, etc.)
- Compiling the scientific and social programs
- Inviting speakers and technicians
- Providing reagents, consumables and some smaller instruments
- All marketing activities (publications, TV and radio reports)
- Printing lectures and exercise books and CD-ROMs
- Updating the website (www.biogentabor.hu)

The University had the following duties:

- Providing lecture room, laboratory and infrastructure

Speakers and technicians were provided mainly by the university and Bio-Science. The number of applicants was three times higher than that could be accepted in both years. Sixteen students participated in each group of the one-week training. All together, 30 kids finished the program in 2003 and 32 in 2004.

## 2. Program details

During the one-week-long scientific program the students got an overview about the molecule of DNA which carries all the genetic information. They got familiar with the basic and the most recent phenomena and ideas. (Table 1.)

**Table 1.** The program of a one-week-long course including lectures and laboratory works.

| Day | Lecture/ Lab work | Subject/Title |
|---|---|---|
| 1st day | Lecture | Science on molecular level in Medicine |
| | Lecture | What is the good researcher like? |
| | Lecture | Cloning, hybridization |
| | Lecture | Labor safety –How shall we work in a molecular biology laboratory? |
| | Lab work | Becoming acquainted with laboratory instruments; practicing of basic techniques |
| 2nd day | Lecture | Diagnostic methods based on molecular biology in medicine |
| | Lecture | Methods of DNA isolation, spectrophotometry |
| | Lecture | PCR- Polymerase Chain Reaction |
| | Lab work | DNA isolation from the students' own blood, measuring of purity and concentration of DNA |
| | Lab work | Examination of SP1 polymorfism of Collagen I. alpha-1 gene with PCR-RFLP technique |
| 3rd day | Lab work | Digestion of PCR product |
| | Lecture | Detection methods in molecular biology: gel electrophoresis, gel documentation systems |
| | Lab work | Preparation of agarose gel |
| | Lab work | Running and evaluation of PCR product, DNA staining, gel documentation |
| | Lecture | Real Time PCR and LightCycler system |
| | Lab work | Examination of mutation of Prothrombin gene causing thrombophilia with kinetic PCR techniques |
| 4th day | Lecture | Morphological methods in genetics |
| | Lecture | Apoptosis, programmed cell death |
| | Lab work | FISH practice |
| | Lab work | Apoptosis, programmed cell death |
| | Lab work | Cytogenetics |
| 5th day | Lab work | Evaluation of kinetic PCR products |
| | Lecture | Protein and gene network in the cell: the strength of weak interactions |
| | Lecture | Genomics |
| | Lecture | DNA microarray technology |
| | Tests | Testing the acquired knowledge of the students and the work of the organizers. Evaluation of the tests as well as the camp. Closure. |

The students spent approximately half of their time in the molecular biology laboratory where the experiments were performed with high-tech instruments. (Figure 1.)

**Figure 1.** High-tech amazed campers.

## 3. Results

Students in the camp were extremely enthusiastic. They were active during the lectures asking a number of appropriate questions. Sometimes the flow of questions had to be disrupted in order to stick to the schedule. However, these discussions were continued in the breaks. The practical part of the program was even more exciting for the students. The fact that they were worked with DNA and genes, looking for mutations, was very intriguing for them. During the first course in 2003, we observed that finding a medically important mutation in the genome of one of the students might impose serious ethical questions. Therefore, we decide to make the students look for polymorphisms carrying low importance in their genome. The significant mutations were shown in unknown samples.

The students could follow the evolvement of methods starting from the purification of nucleic acids through the different polymerase chain reactions to cytogenetics and sequencing of DNA. All of these experiments were carried out by themselves! During the course, we have dealt with the misunderstandings around genetics as well as the ethical questions. The social programs (evening party at one of the main organizers' house, bowling party, etc.) have strengthened the groups as a team and lifted the spirit during the whole course. (Figure 2.)

**Figure 2.** Playing games brings people together.

At the end of each course, the students filled out a test to evaluate the efficacy of the camp (Figure 3.).

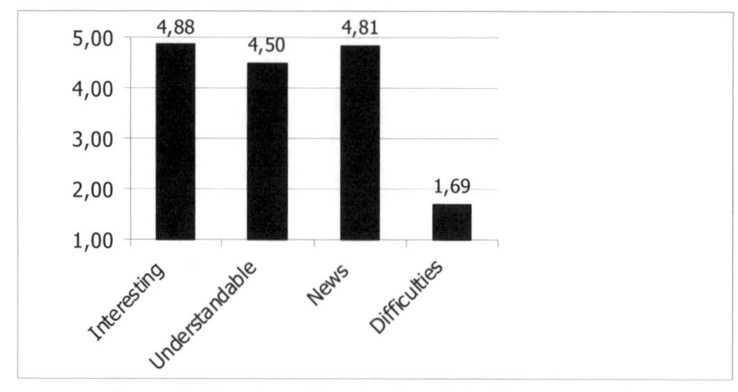


**Figure 3.** How the students evaluated the laboratory work.

Based on our experiences, we consider the BioGen Summer Camp as a successful program (Figure 4.)

**Figure 4.** Happy End.

## 4. Evaluation

The extreme overbooking of our camps indicates the need for such an auxiliary form of education. I believe that it might be important in all areas of education but our selection, genetics was a good choice due to the fast changing characteristic of this field and the public interest in it. The efficacy of our course could be seen not only on the successful results of the written tests but also on some students who have decided to follow a molecular biology track in their further education.

The teaming up of a profit-oriented company and an academic institution appears to be an efficient and successful solution to this type of need in the education. The company can provide the necessary means for the development and organization of such a camp, while the university serves as an ideal location for the students eager to learn. All the teaching tools (CDs, handouts, etc.) have proved to be useful, and some of them have been placed onto the web site of the camp as it was requested by the participants. It should also be mentioned that the students would like to stay in touch with the organizers as well as each other. This is another indication of the success.

We think we have achieved our goal since we managed to contribute to the search for and professional improvement of young high school talents. Therefore, I think that it is in the interest of the society to provide talented students the opportunity to participate in BioGen-like courses in the future. This program can be further developed, more structured, offered at different levels, perhaps including high school science teachers as well in the future.

## Acknowledgements

I would like to thank to Prof. Peter Csermely, Prof. Andras Falus, Prof. Laszlo Kopper, Prof. Tivadar Tulassay, Dr. Laszlo Tancos for their valuable help.

 *Science Education: Best Practices of Research Training for Students under 21*
*P. Csermely et al. (Eds.)*
*IOS Press, 2005*


# Science Festival School: Molecular Biology for Common People

**Joanna LILPOP**
*Science Festival School,*
*4 Trojdena str., 02-109 Warsaw, Poland*
*aska@iimcb.gov.pl*

**Abstract.** A social consciousness of what Science is about is a fundamental need of these times. The Science Festival School (SFS) was founded in October 2002 to educate and popularize modern biology in Polish society, mainly through organizing lectures, workshops and other events for students, teachers and all interested persons. During two years of activity SFS attracted over 1000 young participants in workshops, 2600 people to open lectures and 450 teachers to technical workshops. Over a short period of time SFS has become a success and has been recognized as an equal partner in international educational projects.

## Introduction

Science Festival School (SFS) received its name from the Science Festival in Warsaw. The aim of the Science Festival is to reduce the gap between science and society in Poland. Series of science-related activities, open and free to the public, take place during the two Festival weeks in September. Almost all of Warsaw's sciences, science-related and artistic institutions organize workshops, exhibitions, lectures, various performances and excursions. During the last seven years, the Festival has attracted an increasing amount of people, reaching 60,000 visitors per year.

The founders of the Science Festival School decided to spread the idea and activities of the Festival through the whole year as the first full-time biology popularisation institution in Poland. The International Institute of Molecular and Cell Biology in Warsaw (IIMCB), Institute of Biochemistry and Biophysics PAS (IBB PAS), Nencki Institute of Experimental Biology PAS (NIEB) and the Science Festival itself signed a special agreement on October 1st, 2002 establishing the Science Festival School [1]. Later, The Warsaw Agricultural University joined the founders' committee. Although the Science Festival involves all sciences, the Science Festival School focuses on molecular biology and genetics.

The aims of the SFS as specified in the Agreement are:

- Conducting Educational activities popularizing the theme of biology in Poland by organizing open lectures, workshops for students and all interested people, courses for biology teachers, and exhibitions.
- Improving biology education and awareness of biology in society [1].

## 1. Organization and support

SFS acquires a professionally equipped laboratory and good technical and science support from the founders' institutes and through the supervision of the Science Committee. The School's office and administration are located at IIMCB while laboratories are at IIMCB and Warsaw Agricultural University. Outstanding professors from the founder's institutes serve on the Science Committee: Jerzy Duszyński, Jacek Kuźnicki, Magdalena Fikus, Halina Wędrychowicz, and Włodzimierz Zagórski-Ostoja. The Institutes and sponsors finance the School and supply chemicals and equipment needed in the School's activities. During the school year 2003/2004 SFS's sponsors were: Fermentas, MP Biomedicals, Scie-Plas and Symbios. An important media partner is the biggest Polish daily newspaper, Gazeta Wyborcza (with science popularizing initiatives "Akademia Szkoły z Klasą" and "Wolontariat Studencki" in which SFS takes place). The British Council, Nowa Era publisher, Radio Bis with Science Picnic and DANA Foundation (with International Brain's Week) are regular partners. To be in touch with society, SFS takes part in all events popularizing science, such as the Science Festival, Science Picnic and Celebrating the DNA50' with the British Council. Collaboration with the National Centre for Biotechnology Education at the University of Reading, who are SFS's great friends, has resulted in many successful projects such as the adaptation of the "kitchen laboratory" curricula to Polish textbooks for teachers and schools.

SFS now has several persons, who are scientists from Institutes and Universities and biology students, creating the programme and conducting the workshops.

## 2. Touching the science

SFS during the school year regularly organizes workshops in its professionally equipped laboratories. Participants, who are mostly high-school students, apply laboratory equipment, techniques and real-live experiments. The practical experiments are supported by lectures presenting the theoretical basis of molecular biology, genetics and its techniques. The workshops last about 6 hours during one day and cover topics such as examining DNA by PCR methods, bacterial transformation, gene cloning, protein fingerprinting or molecular diagnosis. The group leaders, always two or three in the room, are available to answer questions and to help in laboratory practice.

Most of the participants are school groups but anyone can participate by making an application and solving an on-line biology test. Since the beginning of this activity, SFS has hosted over 1,000 participants (see table 1.). Among them were groups of Warsaw Technical University students and finalists of Polish Biology Olympiad who advanced to the international final.

Besides teaching laboratory practice, SFS tries to present theoretical issues of modern biology. Every two weeks open lectures on molecular biology are organized, given by top Polish scientists. The lectures are accessible to anyone with a basic knowledge of biology. Some topics popular in media like genomics, evolution controversies, genetic diseases or genetically modified organisms bring huge audiences. But also more specific themes relative to gene expression regulation, immunology and the theory of evolution are in the center of interest for young students and biology teachers. On average, the lectures attract over 100 listeners.

A programme to train biology teachers was started in response to a big interest among the schools. The first four-day course prepared in collaboration with Teachers Excellence Center in Warsaw ended with great applause. The participants not only had a chance to learn how to use modern laboratory equipment and molecular techniques, but also how to make some "kitchen biology experiments" that can be easily implemented in the classroom. Therefore from its second year, SFS started offering such courses regularly. The SFS laboratory entertained 40 teachers from Warsaw high schools and decided to spread the idea to the whole country. Through the support of Gazeta Wyborcza, SFS organized 5 weekend meetings at Teachers Excellence Centers throughout the country. In cooperation with the publisher of Nowa Era, one-day workshops on kitchen laboratory methods were organized in the 10 biggest cities.

Finally SFS was the main organizer of workshops for biology teachers in Warsaw, coordinated by European Molecular Biology Organization. This event was one of nine in Europe financed by the European Union as a part of the project "Continuing Education for European Biology Teachers". A total of 74 teachers from all over the country (21 from Warsaw) and 3 from the Ukraine participated in a four-day workshop, 17-20 June 2004.

To complete these activities designed for teachers, SFS in cooperation with EU Centre of Excellence at NIEB (BRAINS) organizes one-day conferences for teachers each year.

**Table 1.** The summary of SFS activities during the two years.

| **Activities** | **I year 2002/2003** Number of groups / number of participants | **II year 2003/2004** Number of groups / number of participants |
|---|---|---|
| Workshops for youth | | |
| Workshops for youth in SFS | 34 / 500 | 41 / 650 |
| Workshops for Warsaw Technical University students | | 2 / 35 |
| Workshops for biology Olympiad finalists | | 2 / 20 |
| Workshops in primary schools | | 10 / 250 |
| Courses for biology teachers | | |
| Four days courses in SFS | 1 / 16 | 3 / 24 |
| EMBO workshop | | 1 / 77 |
| Weekend courses for secondary school teachers | | 5 / 65 |
| Workshops for teachers co-organized with Nowa Era Publisher | | 10 / 300 |
| Open lectures | | |
| SFS Lectures | 11 / 2000 | 12 / 720 |
| Mini-symposium for teachers with BRAINS | 1 / 50 | 1 / 30 |

As Table 1 shows, the first year was devoted to student workshops primarily targeting the Warsaw area. During the $2^{nd}$ year SFS developed and changed its profile to focus on teachers and to cover the whole country. Training teachers has greater impact and better possibilities for improving biology education in schools.

## 3. The perspectives

From the popularity of the SFS, it is now recognized that such activities are necessary and expected in Polish society. Biology, especially molecular biology, is a quickly developing domain of science. Therefore contact between scientists and society is essential. School education in Poland is still encyclopedic and theoretical, not experimental. But lack of money and education system problems cannot stand in the way of development and improving of biology education. We are obligated to make biology education more interesting, inspiring and to deal with real life.

Attendance at the SFS inspires young students to engage in science and inspires teachers to practice new ways of teaching. The opportunity to make the workshops is rewarding not only for its participants but also for SFS members – young scientists creating and conducting the lessons. It gives life experience, practice in laboratory and schoolwork, and produces professional excellence and great satisfaction.

Through cooperation with institutions such as the British Council, EMBO or NCBE, SFS is open for European and worldwide educational initiatives. Adaptation of their experiences to Polish conditions is not only possible but also necessary for future development of Polish science.

**References:**

[1] Porozumienie założycielskie Szkoły Festiwalu Nauki, Warsaw, 2002

P. Csermely et al. (Eds.)
IOS Press, 2005


# The Romanian Experience within the Comenius Project "Hands-on Science"

**Dan SPOREA and Adelina SPOREA**
*National Institute for Lasers, Plasma and Radiation Physics,*
*409 Atomistilor St., Magurele, RO-76900, Romania*
*sporea@ifin.nipne.ro*

**Abstract**. The paper refers to the main results obtained by the Romanian network of high schools participating to the EU's Comenius project "Hands-on Science": development of teaching software aids; the use of virtual instrumentation in high schools laboratories; the training of students to prepare their own experiments.

## Introduction

One of the many challenges our society is facing in the modern world is the growing discrepancy existing between the rising demand of highly qualified manpower (in the field of science and technology development) and the low number of young, talented people graduating technical schools. This is a concern on both sides of the Atlantic, and governments, companies and professional societies make efforts to overcome this situation. Almost everyone realized now that, in order to have students interested to study science, we have to consider more carefully the problem at all levels of the education pyramid: primary, secondary and high school [1]. The way children are trained to approach science understanding as they grow-up has a major impact on their future options and affinities. In the North America, a consortium built around SPIE, OSA, MESA and NOAO obtained just one year ago an important financial support ($1.7 million grant) from the National Science Foundation in order "*to design and implement an optics-based educational enrichment program for middle school students in the United States*" [2]. Major professional bodies have societies dedicated to teaching activities, and in some cases they have a special focus on the improvement of technology teaching in high schools [3]. The European Union recognized the importance of science literacy by supporting various projects in the frame of Socrates Programme (Leonardo da Vinci, Erasmus, Comenius [4]), an annual contest for young scientists [5], a major conference [6] and a forum for the dissemination of science results among teachers in Europe [7]. In the frame of the EU's funded Comenius project "Hands-on Science", our Institute is managing in Romania, at the national level, a network of very active high schools aiming to attract students towards science and technology study by offering them the possibility to learn and to express themselves through experiments. A complete presentation of the overall project, as it operates at European scale, can be found on the Internet and elsewhere in this volume [8], [9].

## 1. Romanian team objectives

The *main goals* of our team, as part of the "Hands-on Science" European network are:

1. to transform the school (teachers and students) from classical end-users for teaching and training aids into active designers and developers of such materials;
2. to use IT & C technologies as efficient tools for the democratization of science teaching, by open the door to the 'knowledge treasure" to less favoured players and by assisting with funds/ consultancy/ advertising those able to craft an other face to the traditional teaching schemes and to create new, unexpected teaching aids;
3. to promote as much as possible the generous ideas of the project into a large pool of recipients and beneficiaries (students, parents, teachers, central and local authorities, NGOs, companies involved in the teaching process).

In designing the working plan for the Romanian participation to the project we focused on several directions which can bring the maximum benefit both for the project and for the country, on its way to EU integration:

a. to build strategic partnerships with organizations and companies which can assist us to run the project, either through direct financial support or by associating the project name with their image;
b. to support the inclusion of virtual instrumentation programming teaching in high schools;
c. to assist high schools in developing virtual experiments;
d. to encourage high school teachers to train students in developing their own experimental set-ups and training aids;
e. to prepare teaching materials in electronic and multimedia format;
f. to facilitate the access to experiment-based teaching to less favoured groups (minorities, orphan children, students form rural areas);
g. to disseminate the project results though lectures, conferences, communication sessions.

The paper will refer further to the major results we have according to the above mentioned strategy within the eight months since the project started.

## 2. Major players

The coordinator of the Romanian team is the Laboratory for Lasers Metrology and Standardization, part of the National Institute for Lasers, Plasma and Radiation Physics, located near Bucharest. This entity is accredited as a testing laboratory for laser-based products, and its main research activities are related to the evaluation of radiation effects on optoelectronic components and optical fibers, in the frame of the EU's Fusion Programme. Previously, the Laboratory also ran several projects dedicated to vocational training (Leonardo da Vinci and PHARE). Our research background helped us a lot during this Comenius project, both through our access to up-to-date scientific instruments and investigation methods, and by our contact with companies and professional societies, acting as would-be project supporters.

Our two major partners in the "Hands-on Science" project are "Tudor Vianu" High School of Informatics, in Bucharest, and "Stefan Odobjela" Theoretical High School, the former High School of Physics, also in Bucharest.

According to the project rules, we assembled a national network formed by various high schools and vocational schools across the country, entities which participate to the project as associate members. We want to mention just few of the more active ones:

- "Grigore Moisil" Theoretical High School, in Bucharest;
- "Tudor Vladimirescu" Theoretical High School, in Bucharest;
- "Nicolae Balcescu" Theoretical High School, in Medgidia;
- "Casa Corpului Didactic", a structure supporting teaching activities in different parts of the country.

As mentioned earlier, we established co-operation links with several organisations involved in educational activities. For promoting the teaching of Virtual Instrumentation programming in high schools we relay on the Romanian LabVIEW User's Club (Fig. 1) [10], a Web portal located on a server at the "POLITEHNICA" University of Bucharest (PUB), Center for Advanced Technologies - CTANM. CTANM's activity is centred on Virtual Instrumentation and e-Learning applications. After completing around 15 EU funded projects dedicated to high education, educational networking and students exchange [11], CTANM's Executive Director, Associate Professor Tom SAVU, is now acting also as National Instruments Academic Manager for Romania and as e-Learning responsible for PUB. The graphical programming concepts are part of the high-school curricula since year 2000, when a dedicated book [12] was published and approved by the Ministry. After the Romanian Ministry of Education and Research acquired, in 2002, LabVIEW licenses for several hundreds high schools, now a national Virtual Instrumentation strategy is under development in co-operation with National Instruments Corp., Austin, Texas, U.S.A. The strategy, first of its kind in the world, followed this year by a Chinese one at university level, is defining the methodology and quality criteria for teachers' training and is building the framework for advertising and disseminating the teachers' and pupils' achievements. A section dedicated to the high-school field was specially organised during the first edition of the National Conference for Virtual Instrumentation [13]. The strategy is also trying to encourage the development of computerised measurement didactic applications by the teachers and students and is trying to organise a Web portal for the applications exchange.

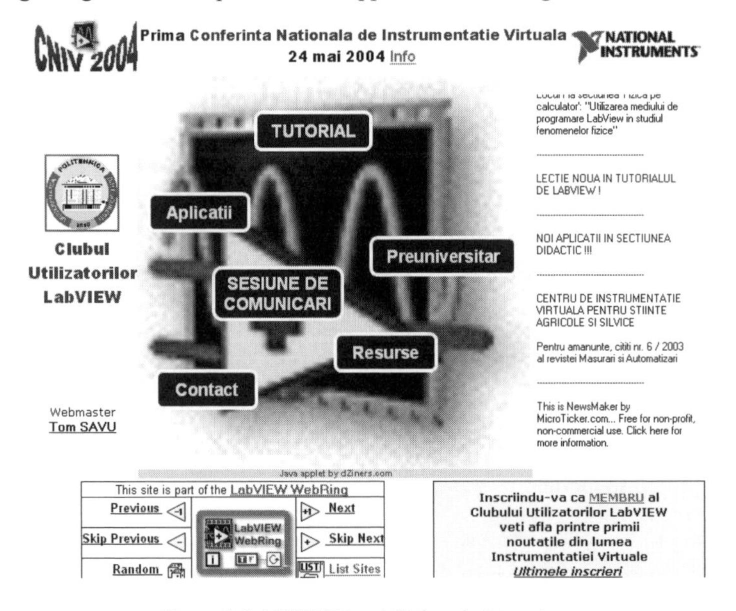


**Figure 1.** LabVIEW Users' Club main Internet page.

Our second partner in this project is the Economic and Administrative Training Centre, an independent e-learning and distance learning training centre, located at the University of Bucharest [14]. Their main field of interest are modern PC-based training courses in IT, administrative studies and law. They are licensed for distance learning teaching. We relay on them, based on their interests and experience, for the organization in Romania of an international workshop on life-long learning, subject following into our field of interest, too.

Another important participant with whom we expect to boost further some of the project tasks is the Centre for Complex Studies, also located at the "Politehnica" University [15]. This Centre activities are animated by Professor Florin Munteanu. They have quite a unique approach to science learning and they are promoting some unconventional teaching methods (a two weeks young student's study camp on an island, research done through small "personal" investigation laboratory which students can use at home, etc.). With the participation of this Centre, "Casa Corpului Didactic" and "Tudor Vladimirescu" High School we expect to reach a special audience for the project: minorities and less favoured students. In the mean time, we are targeting the development in the project frame, of some trans-European activities in order to increase the project consortium cohesion.

## 3. Results

Considering our past experience with graphical programming and the use of virtual instrumentation (LabVIEW and Measure Foundry) for data acquisition and instrumentation control, our Laboratory decided to support, in the frame of the "Hands-on Science" project, the above mentioned programme for high schools teachers training on the use of National Instruments software tools. We developed our approach to this problem in close contact with CTANM, through a common strategy and complementary activities. As the situation differs from high school to high school, our common support concentrated on several different directions [16]:

- For high school already having a PC data acquisition board (there are for the moment very few) dedicated sensors (i.e. force sensor, electro-magnetic field sensors) were bought, which made possible for the teachers to develop some original experimental set-ups, running under computer control. Such an approach is an innovative one for Europe, as far as high school experiments can now become quite complex and very flexible.
- High schools possessing such an interface were also encouraged and assisted by the Dr. Savu to use more trivial sensors (photodiodes, light emitting diodes, etc.) and to prepare some less complicated experiments.
- In the case of another high school a simple PC interface, developed in Romania, was used in connection with LabVIEW programming to collect biological data for human heart activity analysis.
- The vast majority of high schools, involved into this programme and which do not have a PC interface, worked on the development of LabVIEW-based simulation programmes to teach physics through virtual experiments.

Among the most notable results in the above mentioned activities are the experiments prepared at "Grigore Moisil" High School through the efforts of Mrs. Mihaela Garabet and Mr. Ion Neacsu. Their approach was a two-fold one: real life experiments and virtual experiments [17] – [20]. Their work addresses different fields of physics' chapters such as mechanics and electricity. At the beginning they propose some virtual experiments where students familiarize themselves with the problem and simulate various situations. For example, they work with their teachers by investigating basic elements of electric circuits like resistors, batteries, capacitors, inductors, or performed some measurements on the current-voltage characteristics of a bulb or a light emitting diode (LED). In the case of mechanics the teachers try to show to their students the way the oscillator amplitude depends on time, how to evaluate the period of oscillation of a pendulum, or to asses the influence of the medium over the oscillation amplitude changes (Fig. 2). In the second stage, they implement a real experiment, by connecting all the needed components to the PC through a data acquisition interface and gathering real data (Fig. 3). Students are further trained to process the experimental data.

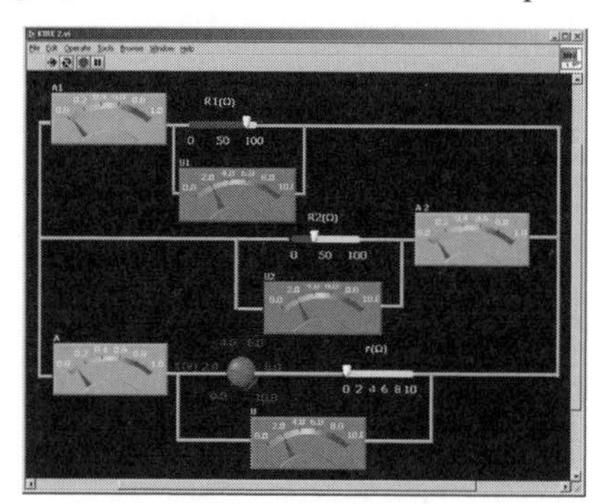

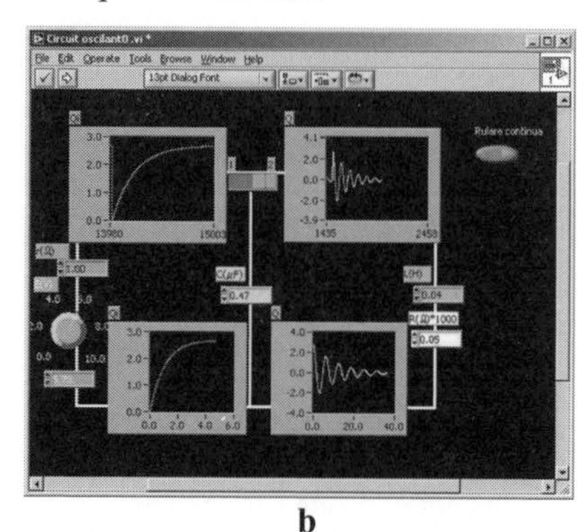

a b

**Figure 2.** Virtual experiment for the teaching of Ohm's law (*a*) and for the study of oscillating circuits (*b*) (courtesy of Mrs. Mihaela Garabet, "Grigore Moisil" Theoretical High School).

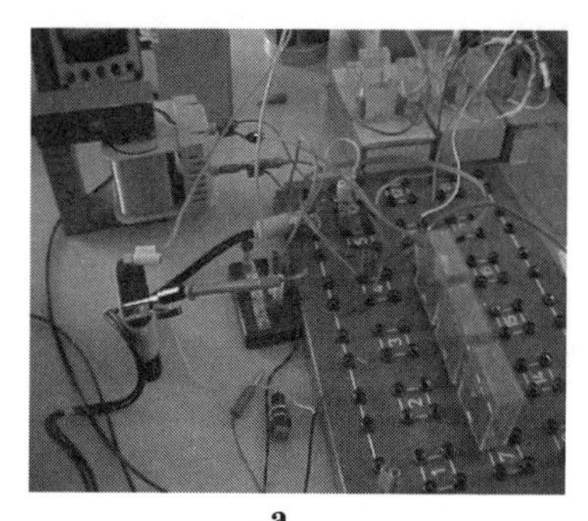

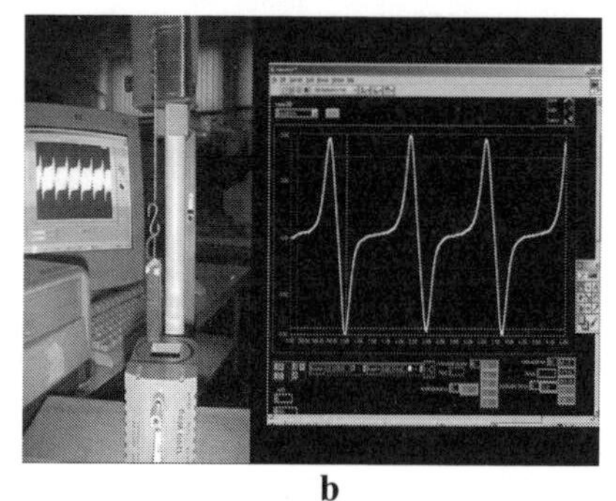

a b

**Figure 3.** Real experiment on the electronic oscillating circuit (*a*) and the mechanical oscillator (*b*), using a data acquisition interface and LabVIEW programming (courtesy of Mrs. Mihaela Garabet, "Grigore Moisil" Theoretical High School).

Interesting results in using LabVIEW to build virtual experiments were reported by Mrs. Emilia Pausan from "Tudor Vladimirescu" Theoretical High School. A remarkable note for her efforts is their diversity. Her work addresses mechanics, optics, and electricity [21]. As an example we mention here the experiments related to the movement of a pendulum (Fig. 4) and the free fall of a body (Fig. 5). Their approach, based on interactivity and didactic strategy, includes also the projection method. Such type of applications could be easily integrated in physic case studies that could be used by a student either at school or during home training.

We have to underline at this point two other innovative aspects of the teaching approach at "Tudor Vladimirescu" Theoretical High School: the implementation of several virtual experiments having 3D representations (i.e. the trajectory of a charged particle entering an uniform magnetic field, or the movement of a conic pendulum), and the tests they developed for students' evaluation. In these tests, the student is asked to build himself a virtual experiment in order to evaluate the values of physical measurands, or to determine the characteristics of a proposed system.

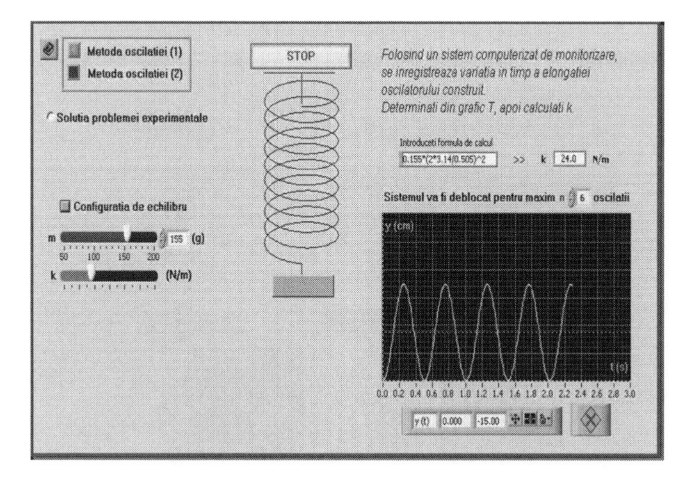


**Figure 4.** Virtual experiment for the investigation of the pendulum movement (courtesy of Mrs. Emilia Pausan, "Tudor Vladimirescu" Theoretical High School).

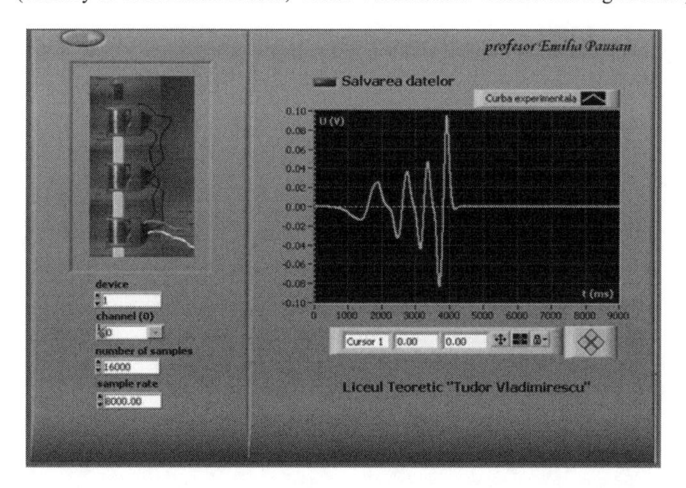


**Figure 5.** Real experiment on the free fall of a body, using a data acquisition interface and LabVIEW programming (courtesy of Mrs. Emilia Pausan, "Tudor Vladimirescu" Theoretical High School).

Some other LabVIEW users focused their work on a specific chapter in physics, as it is the case of Mr. Stefan Grigorescu form "Casa Corpului Didactic" in Bucharest and Mr. Gheorghe Totea form "Nicolae Balcescu" High School in Medgidia.

In the first situation the author developed several virtual instruments to simulate various types of movements of bodies: the *oblique throw* along a direction having an angle with the horizontal plane, the simulation of a *moving body on a inclined plane,* the *straight motion (2D).* Motions are accelerated or decelerated. The author's main concern was to make accessible to young students the basic concepts used in kinematics: trajectory, velocity, acceleration, angle of friction etc. Mr. Grigorescu prepared a set of user interfaces where the student can pre-set the initial motion parameters [22]. After running the programme the results are plotted (Fig. 6) and the motion characteristics are displayed.

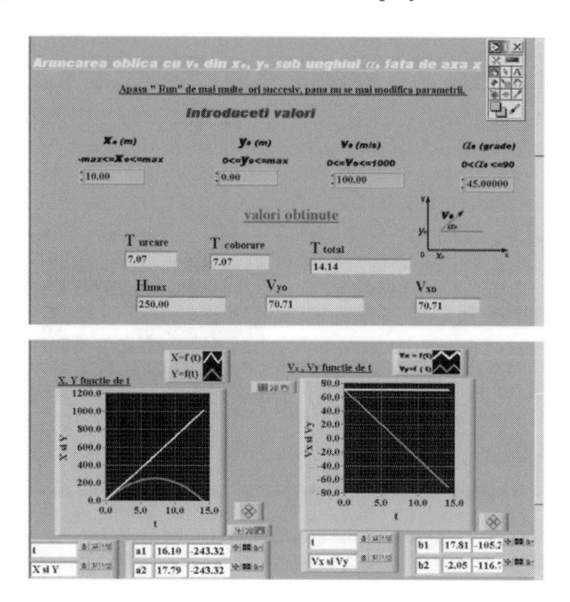


**Figure 6.** Virtual experiment used to demonstrate the movement of a body on a plane (courtesy of Mr. Stefan Grigorescu, "Casa Corpului Didactic").

In the second case we are reporting, Mr. Gheorghe Totea prepared a set of virtual instruments dedicated to the laws of the ideal gas (Boyle's law, Gay Lussac's law and Charles's law) [23]. The suite of programmes can be handled as a entire package and incorporates apart from the experimental section a short historical presentation along with the biography of the scientist who contributed to each discovery. In separate virtual instruments the student can handle in real time various instruments to set-up the experimental input data (Fig. 7). The programme output is given in graphical form making possible an easy and intuitive interpretation of the results. As a novelty, compared to other programmes, the set developed at Medgidia includes in its final part a test used to check the student's understanding of the phenomena he was introduced to (Fig. 8). The test part was also developed using LabVIEW.

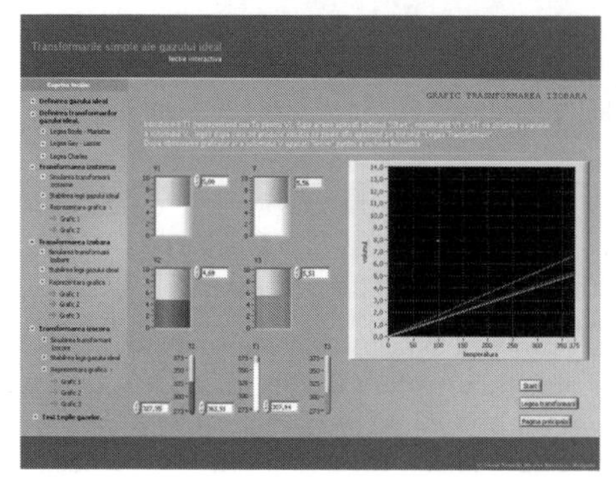
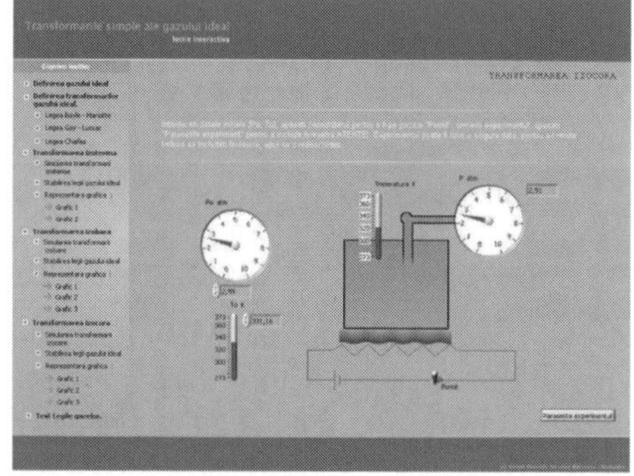

**Figure 7.** Virtual experiment used to demonstrate the laws of the ideal gas (courtesy of Mr. Gheorghe Totea, "Nicolae Balcescu" Theoretical High School).

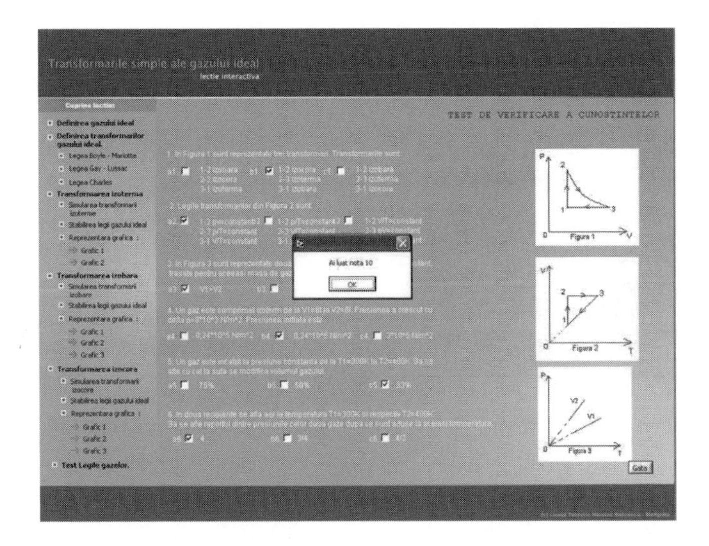


**Figure 8.** The final test of the software package referring to the use of virtual instruments in teaching the laws of ideal gas (courtesy of Mr. Gheorghe Totea, "Nicolae Balcescu" Theoretical High School).

The software aids introduced up to this point were all of them based on the use of National Instruments LabVIEW graphical programming environment. Another direction on the development of IT materials for high school teaching was the preparation of simulation programmes based on the *Macromedia Flash* software. This approach was intensively employed by the team form the High School of Informatics "Tudor Vianu". They contributed a lot with simulation programmes for the teaching of physics [24], [25], informatics [26], [27] and chemistry [28].

Because in the frame of the project we have some very specific target groups with lower financial means (some minorities, rural schools, etc.), where the access to PC-based teaching is not so accessible, we encourage also high school teachers to develop teaching aids based on very simple materials, so that such students can prepare their own small scale experiments. Noticeable results were obtained by Mrs. Marinela Ruset who managed to assemble a group of young enthusiasts to build some very interesting experiments in physics [29]. Under the project's umbrella, a students' scientific session was organised - *I learn better when I build the experiment myself.* Within this frame, several students presented some communications on the *ultrasounds*, the *laser*, the *dioptric telescope*, the *echo-graph*, the *GriMoc manometer*, the *electromagnetic waves*, the *IR light barriers*. A jury rewarded the best students' papers (Fig. 9). The event was accompanied by an exhibition containing teaching materials developed by students (Fig. 10). We were preoccupied also to prepare some tutorials and manuals addressed to high school students. Efforts were made to translate in English, parts of a compendium of definitions and formulae encountered in physics teaching [30]. The work value resides in its use as a memo aid by students preparing to pass the examination to attend university studies.

**Figure 9.** Students demonstrating their experimental set-ups at "Stefan Odobleja" Theoretical High School (from the authors' archives).

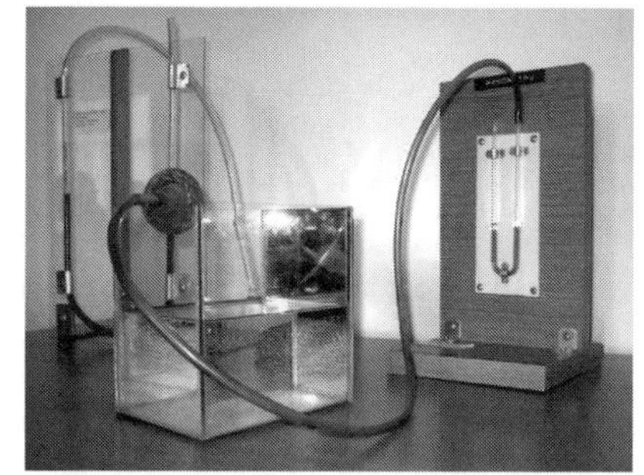

**Figure 10.** Several teaching aids for experiments in physics, exhibited by students at "Stefan Odobleja" Theoretical High School (from the authors' archive).

Another direction we pursued was the development of electronic and multimedia teaching materials. One example refers to a multimedia manual to study physical geography for the first high school grade. Physical Geography was chosen for its distinctive property that it is much easier understood if its teaching implies moving graphics, animation and real-world video. Cumulated with the ability to contain a vast array of graphics, charts and multimedia content, in general, made Physical Geography the obvious choice [31].Visuals include charts and graphs, maps, illustrations, pictures, computer graphics (where a real photo is difficult to obtain or e.g. cross sections are needed), movies and computer animations. These animations depict various phenomena, much like a simulation would do, or allow the student to take a closer look at tools, objects and entities discussed in the lesson. Some pictures are interactive, meaning that clicking hot-spots brings detailed explanations in the definition panel or jumps to the lesson that discusses the respective graphical representation (Fig. 11).

Much effort was dedicated to the promotion of this project at national level and for the dissemination of results. The process of dissemination had two major objectives:

- to make public the Romanian team's achievements;
- to promote the image of the most active participants, as a reward for their endeavour.

Some of the participating high schools shown their results to different national contests dedicated to educational projects. This year the National Conference on Virtual Instrumentation organised by CTANM included a special section devoted to papers presented by high school teachers, developing educational aids using LabVIEW. We can consider that this activity was a real achievement as far as 16 papers were presented. In any case, the result is a premiere for Europe.

**Figure 11.** Original graphics of the solar system in the visual panel: the interactive image reacts to mouse clicks showing relevant details (courtesyof Mr. Radu Sporea, "Politehnica" University - Bucharest).

The greatest success of our continuing efforts is without any doubt the impressive participation of Romanian teachers with papers to the First International Conference organized by the network in Ljubljana. About 40 % of all the conference papers were from Romania.

The promotion of the project goals was done also by an oral contribution to an international conference held in Romania, where the attendees were introduced to the "Hands-on Science" project structure, activities, target objectives [32].

## 4. Conclusions and future plans

After eight months from the project start we can conclude that: the Romanian team succeeded to implement several major themes from the project work-plan, which were selected considering the team strong points, the national interests and the innovative contribution to the project; our major objective - to train young people in *doing science*, was achieved as far as in all projects described in this paper our students *actively participated*; we reached our goal to engage teachers and students in developing training aids as real experiments or virtual laboratories.

For the coming future we have to concentrate our efforts towards: further development of training materials (real experiments supported by PC acquisition and processing; virtual experiments and simulations; e-materials); a higher dedication to interaction with specific target groups; preparation of some activities implying players form partner countries; the organisation of the international workshop on life-long learning.

**References:**


[1] The Royal Society, Statement on the Assessment of Science Learning 14-19, at www.royalsoc.ac.uk/education/assessment.

[2] spie.org/communityServices/StudentsAndEducators/index.cfm?fuseaction=hoo.

[3] www.ewh.ieee.org/soc/es/.

[4] D. Sporea and Adelina Sporea, Comenius "Hands-on Science" Biannual Report, June 2004.

[5] ica.cordis.lu/search/index.cfm?fuseaction=events.simpledocument&EV_RCN=2518&CFID=491862&CFTOKEN=72613083.

[6] ica.cordis.lu/search/index.cfm?fuseaction=events.simpledocument&EV_RCN=15558&CFID=491862&CFTOKEN=72613083.

[7] www.eiroforum.org.

[8] colos1.fri.uni-lj.si/hsci/index.html.

[9] M. F. Costa, The Hands-on Science Project, In: this Volume.

[10] www.ctanm.pub.ro/ClubLV.htm.

[11] www.ctanm.pub.ro.

[12] T. Savu and G. Savu, Informatica – Tehnologii Asistate de Calculator, ALL Publishing House, Bucharest, 2000.

[13] www.ctanm.pub.ro/Club/Prezentari/CNIV%202004/CNIV2004.htm.

[14] www.cpea.ro

[15] www.csc.matco.ro.

[16] T. Savu and D. Sporea, Romanian Strategy on Virtual Instrumentation Training in High Schools. In: Saša Divjak (ed.), Proceedings of the 1 st International Conference on Hand on Science, HSci2004, Teaching and learning Science in the XXI Century, 5-9 July, 2004, Ljubljana.

[17] Mihaela Garabet and I. Neacsu, Understanding the Electric Circuits, In: Saša Divjak (ed.), Proceedings of the 1 st International Conference on Hand on Science, HSci2004, Teaching and learning Science in the XXI Century, 5-9 July, 2004, Ljubljana.

[18] Mihaela Garabet and I. Neacsu, Electronic Oscillators, Between Real and Virtual, In: Saša Divjak (ed.), Proceedings of the 1 st International Conference on Hand on Science, HSci2004, Teaching and learning Science in the XXI Century, 5-9 July, 2004, Ljubljana.

[19] Mihaela Garabet and I. Neacsu, The Motion, Between Real and Virtual, In: Saša Divjak (ed.), Proceedings of the 1 st International Conference on Hand on Science, HSci2004, Teaching and learning Science in the XXI Century, 5-9 July, 2004, Ljubljana.

[20] Mihaela Garabet and I. Neacsu, Mechanical Oscillators, Between Real and Virtual, In: Saša Divjak (ed.), Proceedings of the 1 st International Conference on Hand on Science, HSci2004, Teaching and learning Science in the XXI Century, 5-9 July, 2004, Ljubljana.

[21] Emilia Pausan and Monica Iliescu, Computerized Systems for Physics Laboratories, In: Saša Divjak (ed.), Proceedings of the 1 st International Conference on Hand on Science, HSci2004, Teaching and learning Science in the XXI Century, 5-9 July, 2004, Ljubljana.

[22] Stefan Grigorescu, Experiment simulations (Virtual instrumentation) created using LABVIEW 6.1. In: Saša Divjak (ed.), Proceedings of the 1 st International Conference on Hand on Science, HSci2004, Teaching and learning Science in the XXI Century, 5-9 July, 2004, Ljubljana.

[23] Gh. Totea, Simple ideal gas transformation. In: Saša Divjak (ed.), Proceedings of the 1 st International Conference on Hand on Science, HSci2004, Teaching and learning Science in the XXI Century, 5-9 July, 2004, Ljubljana.

[24] Ioana Stoica and S. Vasilescu, Using Information Technology in the Moder Educational System - The interference of light. In: Saša Divjak (ed.), Proceedings of the 1 st International Conference on Hand on Science, HSci2004, Teaching and learning Science in the XXI Century, 5-9 July, 2004, Ljubljana.

[25] Ioana Stoica, Mechanics: Oscillations. In: Saša Divjak (ed.), Proceedings of the 1 st International Conference on Hand on Science, HSci2004, Teaching and learning Science in the XXI Century, 5-9 July, 2004, Ljubljana.

[26] Corina Achinca, Cecilia Balanescu, Rodica Cerchiu, G. Dita, S. Nistor, Using Information Technology in the Moder Educational System - Backtracking. In: Saša Divjak (ed.), Proceedings of the 1 st International Conference on Hand on Science, HSci2004, Teaching and learning Science in the XXI Century, 5-9 July, 2004, Ljubljana.

[27] Corina Achinca, Cecilia Balanescu, Rodica Cerchiu, G. Dita, S. Nistor, Using Information Technology in the Moder Educational System - Recursivity. In: Saša Divjak (ed.), Proceedings of the 1 st International Conference on Hand on Science, HSci2004, Teaching and learning Science in the XXI Century, 5-9 July, 2004, Ljubljana.

[28] Silvia Moraru, Using Information Technology in the Modern Educational System - Chemistry - ALKANES. In: Saša Divjak (ed.), Proceedings of the 1 st International Conference on Hand on Science, HSci2004, Teaching and learning Science in the XXI Century, 5-9 July, 2004, Ljubljana.

[29] Marilena Ruset. When I am doing myself, I understand. In: Saša Divjak (ed.), Proceedings of the 1 st International Conference on Hand on Science, HSci2004, Teaching and learning Science in the XXI Century, 5-9 July, 2004, Ljubljana.

[30] Marilena Ruset, Physics … in Few Steps. In: Saša Divjak (ed.), Proceedings of the 1 st International Conference on Hand on Science, HSci2004, Teaching and learning Science in the XXI Century, 5-9 July, 2004, Ljubljana.

[31] R. Sporea and M. Matache, Geomedia - Multimedia Manual for the Study of Physical Geography. In: In: Saša Divjak (ed.), Proceedings of the 1 st International Conference on Hand on Science, HSci2004, Teaching and learning Science in the XXI Century, 5-9 July, 2004, Ljubljana.

[32] D. Sporea and M. Costa, Science Teaching in High Schools: the Comenius "Hands-on Science" project. In: Education Facing the Contemporary World Problems, Editura Universitatii din Pitesti, 2004, Vol. II, pp. 57-63.

P. Csermely et al. (Eds.)
IOS Press, 2005


# The 2nd Transylvanian Conference of Students' Research Clubs (TUDEK 2003)

**Ágoston DVORÁCSEK**[1]
*"Bethlen Gábor" College, Aiud, Romania*
*dvoracsek@email.ro*

**Abstract.** The article gives a detailed report of the second Transylvanian regional research student conference (TUDEK 2003) organised in 6 sections and held at the 'Bethlen Gábor' College in Nagyenyed between the 21st and 23rd of November. The organisers and the participants consider it successful hoping that it will become a yearly event.

The Hungarian Research Student Association organised in the autumn of 2000 in Budapest the First National Conference of the Students' Research Clubs (TUDOK 2000) also attended by Transylvanian students. The next year there were so many applicants for the conference that we decided it is better to organise regional conferences in several towns and in the neighbouring countries and that only the prize-winning students from these conferences should compete at TUDOK. The First Transylvanian Students' Research Clubs Conference was organised at Marosvásárhely in the spring of 2002 and the first regional conference from Serbia-Montenegro was held in 2003.

We organised the TUDOK's second Transylvanian regional conference, the TUDEK 2003, at the 'Bethlen Gábor' College in Nagyenyed between the 21st and 23rd of November. We started the organizational work in May 2003. The call for papers was made on the official site of the college (www.bethlengabor.ro), of the TUDEK (www.tudek.home.ro) and also in the central Hungarian newspapers from Romania as well as in some Hungarian county newspapers.

---

[1] Translated by Elemér Fodor, english teacher of the "Titu Maiorescu" College from Aiud, Romania.

Any Hungarian high-school student could compete at the conference (including previous-year graduates). The submitted papers had to meet the following general criteria:

- To have no more than two authors.
- A student or a pair of joint-authors could submit only one paper in a section.
- It wasn't submitted at an earlier TUDOK or TUDEK conference.
- To meet the contemporary scientific requirements (to contain personal observation, surveys or experiments and their interpretations)
- To comply with the formal requirements from the call for papers (without supplements and appendices not to exceed 6 A4 pages, to be written in Times New Roman font types with 12 font size, etc.)

The original budget of the conference was 57,288,000 lei (around 1,400 EUR). We obtained part of the sum from the Communitas[2] Foundation from Kolozsvár, another part from local sources and the remaining part from the Research Student Foundation from Budapest. The board of the college provided lodging for the guest.

The organizing team members from Nagyenyed were Ildikó Stáb, Jenő Krizbai and Ágoston Dvorácsek high-school teachers of the college. They were supported by the board of the college and helped by enthusiastic students.

The Hungarian Research Student Association represented at the Nagyenyed conference by the deputy chairperson Zita Rovó and by the head of the Transylvanian section Lóránd Fülöp from Marosvásárhely coordinated the conference.

96 students from 21 schools from 17 Transylvanian localities submitted 76 papers for the conference. They presented the papers in 6 sections: Biology, Geography-Geology-Environmental protection, other Natural Sciences (like physics-chemistry-mathematics-informatics), Human Sciences (like psychology and social studies), Hungarian Language, Literature and Ethnography. Besides those submitting papers there were seven students who were just curious about such conferences. We could not refuse their request because they are the FUTURE!

As we did not define the fields of research very strictly leaving this up to the students and their coordinators, the papers covered a large area of subjects. It is worth mentioning a few:

- BIOLOGY: the study of ecosystems, medicinal herbs, traditional healing, healthy living and vices etc.
- GEOGRAPHY-GEOLOGY-ENVIRONMENTAL PROTECTION: the protection of mineral wells, the assessment of drinking water, balneology, etc.
- HUMANITIES 1: papers about pedagogy, problems in teaching children affected by the Asperger syndrome, drawings as the mirror of the child's soul, etc.
- HUMANITIES 2: sociological surveys among the gypsies and youth, short monographs about localities, facts about not so well known personalities, etc.
- HUNGARIAN LANGUAGE, LITERATURE AND ETHNOGRAPHY: Károly Kós, the polyhistor, folk diet, traditional trades (wood-carving, furniture painting etc.)

---

[2] With the sum received from the Communitas Foundation we succeeded covering the expenses concerning the prizes for the students, the jury chairmen from Kolozsvár and a part of the expenses with the Hungarian guests.

It is worth noticing that there were more papers, 42 targeting human studies whereas there were only 34 for the science sections. Another interesting fact is that 62 girls submitted papers whereas boys were only 35. Most of the papers, 48 were written in schools from small towns, 24 were written in cities, and only 4 were written in village schools.

The papers were assessed by professors from the "Babes-Bolyai" University and the Reformed Theology from Kolozsvár and then they were the chairmen of the three-member[3] section juries which assessed the presentation of the papers.

The assessment of the papers happened in two stages. The papers were first assessed by the chairmen of the juries from 0 to 20 points taking into account their scientific value, originality and form. During the conference the authors had a ten minute presentation assessed by each member of the section juries from 0 to 10 points. The final grade was the sum of the two separate grades.

I would like to quote two of the jury chairmen regarding the quality of the papers:

*„The papers show the students' interest in novelty related to studying, researching and discovering. For reaching their goals they can make use of what they have learned at school, they study the related special literature and more they conduct their own research."*[4]

and

*„The papers cover a large viriety of subjects, they generally meet the formal criteria, sometimes they missuse some of the technical terms [...] these occasions are very for starting scientific research and preparing professional accounts."*[5]

There were six first, eight second, six third prizes and thirteen special mentions at the conference. The prize giving festivity was attended by the chief editors of the Természet Vilaga and Élet és Tudomány, Gyula Staar and János Herczeg, who presented their magazines and professor László Vekerdi delivered an interesting lecture about Galilei. Thanks to the chief editors the winners received a lot of magazines besides their otherwise small prizes. The winners could also compete in the Fourth National Conference of the Students' Research Clubs (TUDEK 2004) organised at Szeged between 16th and 17th of April where some of them were successful.

We handed out some questionnaires to receive feedback regarding the conference. We learned that there were some problems with the heating, amplification systems, computers, the fact that the jury asked few questions after the presentation and the students needed some overhead projectors! These remarks may be used by the organisers of the next TUDEK which will be held at Barót, Kovászna County in September 2004. It felt good to receive the following remark:

*"In my opinion the event was very well organized and the problems we encountered were due to the poor funding and of the poor technical background but the organisers have done everything that was in their power to make this event a successful one."*[6]

---

[3] Besides the chairman, there were also two secondary-school teachers

[4] Dr. Gábor Pándi, the chairman of the jury concerned with the GEOGRAPHY-GEOLOGY- ENVIRONMENTAL PROTECTION section.

[5] Dr. László Fodorpataki the chairman of the jury concerned with BIOLOGY section.

[6] Szabolcs Vita, a student from Marosvásárhely.

The TUDOK conferences and its regional conferences offer a good opportunity for the good students to present their achievements and to learn about the work of their mates. The students and their teachers can share their experience and speak about their future plans. It would be useful organizing this scientific movement in Romania, too. So far we were not successful because we did not find the proper supporters. Probably, we should start by inviting Romanian students to such conferences. Therefore it would be useful to make a few sections at the next TUDEK where the Romanian and Hungarian students can present their papers for example in English. We hope that we will soon have the financial and the technical support which will ease the work of the organizers because such conferences will be needed in the future, too.

*Science Education: Best Practices of Research Training for Students under 21*
*P. Csermely et al. (Eds.)*
*IOS Press, 2005*


# The Student Research Movement in Vojvodina

**Zoltán KAZINCZY**
*Intercultural Youth Center,*
*Vojvodjanskih brigada 17, Novi Sad, Serbia and Montenegro*
*ifikozpont@neobee.net*

**Abstract**. The most important purpose of the School Research Movement is to give a chance to develop for those young people who are better than the average at certain skills. Within the framework of the program an important role is attributed to forming good teacher-student relationships. It is important to arouse the desire of the young for doing research, which later can result in greater interest, deeper learning, and more concern in work. The School Research Movement would like to facilitate both the autonomous way of thinking of the young at schools and also their independent activities. The School Research Movement has not achieved a lot so far, since it has been existing only for a year in Vojvodina. The efforts made right up to now support only the plans for the future.

## Introduction

It is the bigger cities that are the centres of institutionalised scientific activity in Vojvodina. The opportunity of doing scientific research presents itself only for those young people from the countryside who have financial support besides good school achievement and talent to attend higher education.

The most important purpose of the Student Research Movement is to give a chance to develop for those young people who are better than the average at certain skills. Moreover, another aim of the movement is to arouse the interest of the young at the earliest age possible to do scientific work and to become immersed in sciences. We would like to facilitate both the autonomous way of thinking of the young at schools and also their independent activities.

On the other hand, there is serious need to motivate teachers and professors to initiate such organized activities, to pay more attention to talented students and to be their mentors if they are doing some research.

One of the greatest advantages of the movement is that it connects the student with his mentor through the Internet where there are no borders or differences between villages and cities any more. This way the richness of fields offered by the mentors becomes more attractive for them.

Within the framework of the programme an important role is attributed to forming good teacher-student relationships. It is important to arouse the desire of the young for doing research, which later can result in greater interest, deeper learning, and more concern in work. This reaction will make the social activity of the young more intensive. Those young people who are busy with creative work, researches or sports are not likely to become addicted to drugs, other stimulants or even become members of a sect.

Another important basic principle of the programme is to give opportunity to the young to feel like adults and at the same time they have to be persuaded to behave like that (for instance in a mentor-student relationship). Any research presupposes such independence that a ten-year-old child cannot possess. This independence is challenged when the student has to apply for the job on his own.

The Research Student Movement is on the way to expand the network of the students in Vojvodina who do researches. This programme spreads out to other countries and is bound to the work of similar movements in nearby countries.

Another purpose is to reward those students morally or in other ways who can claim to have achieved something. This can be a good motivation for the future development of the young.

The Research Student Movement in Vojvodina deals with secondary school students as well as with university students. We give opportunity to all the students who join the movement that after they have finished secondary school they can go on doing their unfinished researches.

Our common experience is that if a student joins the programme that depends to a great extent on the attitude of the teachers working with them and also on the attitude of other students of the same age.

In those schools where there are no technical or other special groups, there is no motivation for the students to work independently on a research.

Since the establishment of The Research Student Movement it has changed so much that the students can be informed through several media about the success of their mates, which can have a stimulating effect on their self-confidence and want for work.

From the possibilities that we have at our disposal The Research Student Movement in Vojvodina has put emphasis on the following activities:

**1**. The list of the mentors in Vojvodina:

In order to compile the data of the list we look for professors, teachers of special subjects and researchers in laboratories who want to participate in researches together with students.

Here we collect the data of the professors and teachers who will be our partners. We systematize these data and later the students can choose mentors for themselves from the list.

The systematized data contain where and how the teachers are available, the fields in which they work and the address of the institution where they do their researches.

**2**. Permanent Internet connection:

This makes it possible for students to learn news, or they can give information by helping others. Teachers are able to send useful data, or announce a competition on the Internet.

**3**. Educational lectures:

We would organize these popular-science lectures by involving students who have already been participants at Scientific Student Conferences. These lectures would be given in the schools of 12 towns in Vojvodina where there are Hungarian classes. The aim of these lectures will be to give direct information about the possibilities, jobs, mentors, and the Scientific Student Conferences, etc… to students, so their interest can be immediately aroused for researching.

**4**. Days for Student Researchers:

We would like to organize these days whenever there are youth festivals in Vojvodina. During these lectures, we would turn our attention to the questions that come up in connection with secondary school students (contents of essays, form). The number of the lectures given is limited at scientific organizations, that is why, the programme requires the preparation of students for making posters. At this point, it is important that they could express their feelings with full of ideas, so that it can attract attention. The self-confidence of the students must be supported in order to take part in the work of study groups more actively. They must be informed that anybody may become a researcher, there are no ideal researchers, and it is not always the best students that will turn out to be to the best experts.

**5**. A Conference for The Scientific Conference for Secondary School Students:

The appearance of the young at the Scientific Student Conference for Secondary School Students helps the development of both their personality and identity in many ways.

The aim of the Scientific Student Conference for Secondary School Students is to help stimulate the independent scientific researches of the young researchers', to help creatively study certain scientific fields, to give possibility to students for presenting and getting acknowledged their results. Furthermore, they can compete among one another.

The Scientific Student Conference for Secondary School Students is about to become an important factor in the education of the generation of young researchers. It gives a general survey about the talented and active young people and makes it possible to collect their advantages and disadvantages.

Such conferences give possibilities to give presentations, to make friends, and to gain experience. Besides all these things, these conferences establish a peculiar professional discussion between students, teachers, and the public opinion, without which it is impossible to keep up with the international scientific life of the 21st century. At the conference, there will be students as guests from nearby countries who are also members of similar institutions in their countries and they will perform the results of their projects.

It must be emphasized that the major part of the students in Vojvodina do not study their trades in their mother tongues that is why we have to devote great care to their learning jargons in their mother tongue in order not to remain at a kind of a pidgin language level. This organization gives an excellent opportunity to appraise the jargon of the national minorities and to practice it in everyday situations.

All the essays, projects that were handed in by the members of the conference had been successfully defended. These will be published as scientific proceedings by the Intercultural Youth Centre in a Conference Volume in which the works will be summarized in an extract that will be written in a foreign language. The Scientific Student Conference for Secondary School Students helps the young researchers with their first publications.

This way the publications will be parts of application forms for scholarships, research projects, and study trips, and finally it will be an important part also in the CVs when applying for a job. The successful performances can also be listed in CVs.

**6**. A Conference for The Hungarian Scientific Student Conference in Vojvodina:

In the frameworks of the Hungarian College for Higher Education in Vojvodina get the opportunity to present their scientific works, or the results of their researches in their mother tongue.

The organizer of this Conference is the Hungarian College for Higher Education in Vojvodina that has been considering the institutionalisation of the tradition of the scientific student circle activity as one of its basic purpose ever since it was established.

**References:**

[1] Peter Csermely, Kutatási lehetõségek középiskolások részére, Mũszaki könyvkiadó, Budapest, 2003.

P. Csermely et al. (Eds.)
IOS Press, 2005


# Petnica Science Center: Model of Intensive Extracurricular Science Education of Gifted Students

**Vigor MAJIĆ**
*Petnica Science Center,*
*POB 6, 14104 Valjevo, Serbia*
*vigor@psc.ac.yu*

**Abstract**. Petnica Science Center has started about twenty years ago as an independent experiment initiated by a group of students and science teachers in order to demonstrate a new model of organizing students in science and research. In a couple of years it became a strong center of extracurricular science & technology education of gifted children in former Yugoslavia and an active source of promotion of science and improving sci+tech education in hundreds of public schools in the country. In spite of all terrible social, political, and economical circumstances, Petnica Center survived and developed many innovative methods and models of gifted education used by teachers in a number of schools in Serbia, Montenegro, Bosnia and Herzegovina, Macedonia. Each year, Petnica Center organizes about 150 workshops, science camps, and seminars for 1,500 students and 1,000 teachers from about 500 schools covering a wide spectrum of sciences, technologies, and humanities.

## Introduction

The Petnica Science Center is an independent non-profit organization focused on extracurricular parallel-to-school science and technology education of gifted and highly motivated children and students in age range 13-20 (grades 7-12 + first years of university studies). It is also a number one national in-service teacher training facility in sciences and humanities. As an independent institution, the Petnica Center is free to implement fresh and up-to-date facts, ideas, and methods regardless to fluctuations in national education policy. Educational activities in Petnica cover a wide spectrum of fields of sciences, humanities, and technologies. In the course of each and every Petnica program, reasoning and observation skills, data collecting techniques, argumentation and communication skills are developed. PSC encourages participants to think free, to co-operate with partners from different regions and cultural background. More than 1,000 scientists, engineers, educators, and managers from Serbia and abroad, are included in designing such programs. Today it is one of the biggest system in SE Europe for recruitment of young people with an aptitude in reasoning skills. The main goals of this program are:

i. To identify gifted secondary-school and university students from all parts of country, especially from provincial and poor regions and give them intensive individualized extracurricular education regardless of their ethnic origin, gender, and regardless of religious, or political affiliation of their parents;
ii. To enable the best students to do scientific projects based on real problems and carried out on professional scientific equipment and under the supervision of the best scientists and science teachers;
iii. To instruct young science teachers on how to apply up-to-date scientific concepts, knowledge and educational methods;
iv. To initiate cooperation and exchange of knowledge, experiences and ideas among undergraduates and graduates who study at different universities and different programs;
v. To establish rich international and intercultural contacts and cooperation among young people, students, and teachers.

## 1. Selection Procedure

The most of annual programs and courses are part of Petnica's "Annual Cycle", i.e. a system of complement training activities for secondary-school students (grade 9-12). The first group of courses are "Winter courses", seven day long and organized in February and March, just after the winter holidays. In order to be invited and to take part in these courses, students have to pass complex selection procedure.

Each year, at the beginning of the new academic year, in September and October, Petnica Center sends letters to about 500 secondary schools with detailed information, instructions, and forms. It is up to schools to hang this information at some public place and to inform their students about the programs and applying criteria. Deadline for applications is end of November. In the first part of December a team of about 50 people who cover all fields of Petnica's science education activities, come in Petnica and stay about three days in order to carefully study each application form and attached documentation. The complete documentation from each candidate has to be analyzed by four different Selection Committee member. After that, a final list of participants of the following courses is completed and they are invited to come and take part in some of about 28 week-long courses.

Selection is based on six major elements:

1. Estimation of student's level of motivation and interest for the specific area(s) of science and knowledge,
2. Estimation of student's background knowledge and experience,
3. Individual extracurricular activities, hobbies, and homework,
4. Elements in the application form that can connect above values with the living+learning conditions in applicant's school and family,
5. Recommendations of school teachers and school psychologist, results on official psychological tests if the school practices to use them (it is not necessarily),
6. Impressions about candidate's general literacy, vocabulary, and general fields of interest.

Each year (based on statistics 1995-2003) Petnica's Selection Committee has to study documentation of about 1,500 applicants and to finally select about 600 of them.

However, selection process is not completed here. It is, in some other forms, continual activity. After Winter courses, participant – about 750 students (600 "freshmen" selected by Selection Committee, plus about 150 "old participants", i.e. students who took part in Petnica's courses and camps in previous year), are faced with second selection – some 500 of them will be invited to participate in Spring Courses, and about 300-350 of them on Summer Science Camps.

The number is continually reduced. The core of the most successfull participants (according to quality of their science projects, expressed motivation and ability for individual work) have some special privileges, e.g. they can come in Petnica almost any time and use almost all equipment, books, chemicals, etc. for experiments or some other small projects.

## 2. Effects

Summer Camps are the core programs where participants are in position to do small research projects in science and humanities. They are free to design their own individual or team project and to go through a complete process of scientific research including the final report and scientific paper and presentation of results at Annual Science Conference.

Through "Petnica Annual Cycle", a student can participate in 4-5 courses/camps (in total about 30-34 days). In the following year he/she can apply again passing through much simpler procedure. But, it is expected that he/she will attend different types of programs and prepare more complex scientific project.

After 2-3 years of attending Petnica's programs, some students are well prepared for real and professional science training at university. Among other things, they are trained to use various information sources (books, journals, databases, internet sources, etc.), to communicate with professional scientists and university teachers, to design small research activities, etc.

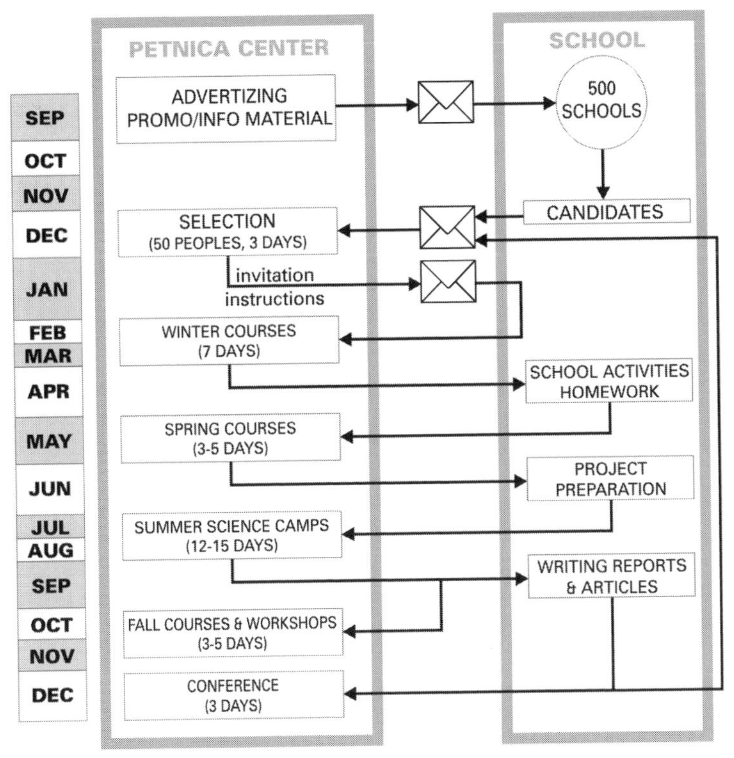


**Figure 1.** Annual cycle of Petnica Center's educational programs. After attending various types of complement programs student can apply again for the programs in following year. He-she is free to attend programs in the same, similar, or completely different area.

## 3. Hidden Goals

Contrary to some common beliefs, the main goal of the Petnica Center is not to prepare future scientists. Petnica Center is devoted to answer to the needs and expectations of gifted boys and girls to upgrade knowledge beyond limits of official school curriculum. Petnica's ethical codex insists that their further professional career is up to them. Petnica Center sees its results or outputs equally successfull if its participants become excellent professional scientists or if they decide to leave science and look for other career challenges. If they are truly gifted, they must be competent and qualified to make free decisions relate to their future. For example, if everybody who likes astronomy becomes astronomer, astronomy as scientific profession will fall in trouble, because astronomy is dependant of many other professions such as journalist (to rise public awareness), teachers (to prepare new generations of scientists), politicians and businessmen (to assure adequate financial and legal support), etc. This is a reason why Petnica Center sees its role to support everybody who is gifted and likes some kind of science, even in case that it is just because of intellectual curiosity or hobby.

Another special goal is to improve students' thinking skills and abilities for general types of critical and creative thinking. It is especially important keeping in mind the recent social and political tensions and conflict within the region. Petnica see its responsibility to help young people to become tolerant, flexible, communicative persons ready to accept democratic values of modern global society.

## 4. Impact on Schools

For enthusiastic teachers who are curious about innovations, sending best students to attend other types of educational programs and extracurricular activities is always a valuable opportunity. They try to implement something new and they ask students to explain what they did in Petnica and, even more, to repeat some of research projects and activities. This is the main reason why Petnica Center regularly publish the most interesting students' projects. "Students' Collected Papers" is published annually in 700 copies and distributed free-of-charge to the most of secondary schools in Serbia and nearby countries.

Apart of students' training programs, Petnica Center organizes a spectrum of in-service teacher training activities. Because of limited capacity, each year Petnica Center organizes between 16 and 35 teacher training courses 3-5 day long. The topics of these courses are innovations in teaching methods, news from science and technology, and exchange of experience in gifted identification and education within existing school curricula.

Now, Petnica Center is number one national in-service teacher training facility. In many cases teachers and school administrators come in Petnica to discuss variety of their problems even beyond the program structure and common functions of Petnica Center. They believe that such experienced and totally independent institution can help them to find good answers to some sensitive and complex problems.

## 5. Can Petnica Center Substitute Regular School Education?

No. There are many reasons why we believe that teenagers must attend regular schools and enhance and enrich their knowledge and experience only through short-time extracurricular programs. One of the top reasons is that young people need to experience rich social life and establish relations with other young people of different interests, different talents, and different experiences. There are some other reasons including the fact that for some gifted boys and girls of age 15 or 16 it is not easy to leave their families without some kind of stress. Of course, there is financial reasons related with the expenses of attending distant school.

## 6. Can Improved School Make Petnica Center Obsolete?

For the similar but complementary reasons no. Well designed educational reform processes could improve the quality of general education including better approach to the gifted and talented students. But the school alone cannot obtain critical concentration of good teachers and good facilities for top 1% of students who need much more beyond existing curriculum. These students need to meet other young people with similar expectations to share experience and ideas. Some kind of regional-level institutions are necessary for good gifted care. Moreover, we can say that the most countries and regions have not enough money to equip all schools with good facilities and good teachers to handle appropriate gifted programs not only in sciences and technologies, but also in sport, art, and other demanding fields.

## 7. Future Expectations

Supposing that the Petnica Center will survive long-term economic crisis in Serbia, it will expand its activities in two strategic directions:

- To improve better multi-disciplinary contents of its educational programs and to make facilities for more complex and demanding students' science projects;
- To develop in-service and pre-service teacher training programs with increasing the number of participating teachers and schools.

Petnica Center has no ambition to expand its programs toward sports and art, because it is not good to have everything within one institution. For Serbia as well as for the entire Region it will be much better to make more specialized institutions in various fields of gifted education. In any case, we see many reasons why such institutions must be of non-governmental type with enough freedom and flexibility to offer truly good programs attractive for gifted young people.

## 8. Conclusion

Petnica Science Center is interesting, unique, complex, but successfull experiment, for many reasons the biggest within the SE Europe. In spite of terrible economic, social, political situation in Serbia, including the war in the former Yugoslavia and eight years of total isolation of the country, it demonstrated that "the gifted can and must survive". The concept "gifted to the gifted", meaning that the most gifted and young participants are involved in design and practicing educational activities for new generations, proved that such approach could produce excellent results and that such institution must stay independent and free of (common) fluctuations and turbulences in governmental policy and priorities.

The region of SE Europe is in need for successfull stories and good continual projects, especially in education where frequent school reforms and reforms of reforme make quakes and sense of instability with no long-term strategy. The twenty-year long experience of Petnica Center could encourages new fresh initiatives and many enthusiastic creative people who already demonstrated the awareness of the role and importance of gifted young people.

*Science Education: Best Practices of Research Training for Students under 21*
*P. Csermely et al. (Eds.)*
*IOS Press, 2005*


# Recruitment of Talents for Life Sciences in Slovakia: State of the Art

**Lubomir TOMASKA**
*Department of Genetics, Comenius University, Faculty of Natural Sciences, Mlynska dolina B-1, 842 15 Bratislava, Slovak Republic*
*tomaska@fns.uniba.sk*

**Abstract**. A general problem of decrease in interest of high school students for science education in Slovakia is amplified by local-specific 'phenotypes' including culturally inherited preference of conformity and discrimination of originality and talent [1]. In this context it is not surprising that Slovak universities are totally passive towards talented high school students. This lack of an assertive attitude might eventually lead to a massive exodus of gifted young individuals and the Slovak universities will be left out with average and marginal importance. To avoid this pessimistic outcome, the implementation of a systematic search for talents, their recruitment and motivation, is a necessity.

## 1. Life sciences and their support and training in Slovakia

Contemporary science is characterized by two major trends. First, there is an enormous increase in the amount of data due to modern powerful technologies. This leads to a demand for original approaches to the novel type scientific questions. On the other hand, there is a general decrease in interest of high-school students for sciences [2]. The situation in Slovakia is even more complicated. Although support of science, technology and education is declared as a political priority, the real investments that would be based upon quality measures are almost negligible. The high numbers (several billion Slovak crowns) presented by administration officials to demonstrate sufficient support of science are totally misleading: most of the resources are allocated by non-competitive means, and their accessibility is limited to a small population of "celebrities", with direct connections to the government. Here is an example of a principle of governmental support of Slovak science and research training at the universities.

If you are a life scientist in Slovakia and you want to apply for governmental support, you have two options. First, you can submit a standard grant proposal to two major grant agencies and wait for an outcome of a peer-review process. As most of the proposals are getting funded, the total budget is diluted into amounts sufficient for 'basic laboratory metabolism', but totally insufficient for either introducing new technologies or at least upgrading an existing infrastructure. However, there is an alternative possibility yielding support by several higher orders of magnitude. The 'protocol' is simple: first you need to become a *local scientific celebrity*, with the personal connection to the influential people in the government. Then you are eligible to assist in the implementation of the program in a fashionable research area (FRA) and write few lines in the bureaucratic newspeak indicating that investment in the FRA is the best way to satisfy the stylish trends. You are a celebrity, so your opinion does not need to be confronted with other members of scientific community. The government decides that several billion Slovak crowns (10-times more than the overall resources from the grant agencies) will be spent on the FRA. You contact a few of your friends and ask them to (i) prepare a formal 'call' for proposals and (ii) keep it secret. As your friends are writing the 'call' and thus know the 'rules', they can simultaneously write the proposal itself. All you need to do is to set up a deadline for submissions that is impossible to reach by outsiders. As a result, the government will receive a single proposal, which is (naturally) funded.

The bottom line is that the trend in the investments of the Slovak government in science and education may appear optimistic on paper, but in reality the administration sticks to the stereotypes inherited from the communistic system. Egalitarism in obtaining competitive (and negligible) funds and corruption in getting access to the large-scale resources catalyzes the progression of the current misery of the Slovak higher education.

Understanding the causes of this pessimistic set-up is crucial for solving the current problems with motivation of young Slovak talents for science education. As exhaustively analyzed by Ladislav Kovac in his numerous studies [*e.g.* Ref. 1], the roots of the average that infiltrates Slovak educational institutions go more than two centuries back: „The cult of egalitarism is deeply rooted in our history." [...] „...pseudodemocratic zeal, paranoid nationalism and xenophobia [...] prevented the development of intelectual aristocracy". According to Kovac, these are the major reasons of the conformity, lack of originality and nurturing of average at Slovak universities.

## 2. Nurturing talents at the Slovak universities: Is anybody interested?

Based on the above it is not surprising that systematic search and nurturing talents in (especially life) sciences is practically absent at the university level. Talented and motivated high school graduates are therefore either diluted into a mass of average undergraduate students or they move to westward (more than 10,000 Slovaks study abroad; only 1,500 foreign students were registered at the Slovak universities). The ignorance concerning talented and motivated individuals is taking place in a *picturesque* context: during the past 10 years Slovakia gained the European (and perhaps the world) record: there is almost 20 'universities' per 5 million inhabitants. Putting aside the fact that majority of these institutions by no means meet the criteria of a university, this 'mass demon' also applies to evaluation of the quality of the individual schools; the main measure of the quality of teaching is the number of 'served' students, not the quality of their education. A long-term problem of Slovak culture described by Kovac – cultivation of average, and discrediting the excellence – is inevitably affecting the next generation of people with a higher education.

The above background is certainly superficial, but it might be helpful in understanding some of the causes of the current phenotype of Slovak universities: a complete lack of interest in high school talents and a total absence of assertivity towards young motivated individuals. While in some cultures teaching and directing the research interests of students represent essential and gratifying aspects of the members of academic community, teaching and one-on-one interactions with selected young individuals is at Slovak universities often considered to be more as a bargain than a means for personal satisfaction.

## 3. Times A-Changin': Are they?

"These are some of the reasons why, seven years ago, I initiated a programme in Hungary to help talented and motivated students between the ages 14 and 20 to obtain first-hand experience of scientific research in Hungarian universities and research institutes. The idea immediately gained overwhelmingly positive response from the Hungarian scientific community", wrote Péter Csermely for *EMBO Reports* in 2003 [3]. As outlined above, an analogous activity in Slovakia is greatly needed. The question is, if the prospect of establishing a formal body organizing a similar program in Slovakia is realistic. Although I would not expect an 'overwhelmingly positive response', there are several reasons for optimism.

First, in contrast to the universities, there are sparks of active approach towards talented students at the level of grammar schools and high schools. Several types of Olympiads (in chemistry, physics, mathematics or biology) are organized at the national level and the winners at the Slovak level are performing reasonably well at international meetings. The grammar and high schools teachers are often assisted by their colleagues from universities. An excellent example of a congenial cooperation of this sort is the long-term activity of mathematician Milan Hejny and his co-workers, whose effort resulted not only in recruitment of exceptional students for Comenius University, but also in numerous unique didactic methods of general importance [*e.g.* Ref. 4] (Paradigmatically, the 'grey' atmosphere at the Slovak universities, led Hejny to a decision to abandon the Slovak educational system).

Second, in spite of numerous administrative obstacles Slovakia has several schools for exceptionally gifted children. The first was established in 1990s in Bratislava, where it survived non-favorable local conditions and nowadays it represents a unique experimental design for education of individuals exhibiting unusual skills [*e.g.* Ref. 5]. In two years, the school will produce its first graduates, who will have to make a decision, where to gain their higher education. A big challenge for Slovak universities is to prevent their exodus out of their reach.

Third, there are groups of people scattered throughout the academic environment, who realize possible benefits of an organization that would catalyze search, motivation, recruitment and career development of talented high school students. People, who know that "...conveying knowledge and an appreciation of science to young people and possibly inspiring some of their students will pursue their own science careers" [6]. Many of them already have positive experience with interactions with high school students, who have spent a few hours a week in their laboratory, then became their undergraduate and graduate students and one of the best co-workers. A good example of a high school frequently interacting with universities is the International Baccalaureate (IB) gymnasium of Juraj Hronec in Bratislava, whose graduates majoring in science must undertake a small experimental project. Several such students, originally tempted to apply for a Slovak medical school or not to stay in Slovakia at all, changed their minds and stayed in the field of life sciences.

## 4. Building a network for talent recruitment in Slovakia: A primer

The above example illustrates the power of close personal interactions for a life-time decision of a young individual. The main problem is that such interactions are taking place mostly by chance, further emphasizing the need for a systematic approach. One such effort aimed at biomedical students just started at two Departments (Genetics and Biochemistry) of the Faculty of Natural Sciences (FNS), Comenius University in Bratislava. The ultimate goals of the project overlap with the current state of the Hungarian programme and include:

1. Organization of a series of workshops in the area of biomedical sciences. The workshops would serve as the 'entry points' for students interested in biology or chemistry into scientific laboratories.
2. Building a network of research groups that would (i) participate in the organization of the workshops, (ii) select suitable students, (ii), provide a database of experimental projects and (iv) provide the students an opportunity to work in their laboratories. The leaders of the groups would also serve as tutors guiding students' decisions related to their future careers.
3. Integration of the Slovak network into its European counterpart, leading to exchange of both students and ideas.

Naturally, we are eager to be at a level of such a large-scale network (Fig. 1) as soon as possible. Taking into account the current situation in Slovakia described in the previous sections, it seems that one scenario for building a network of individuals and institutions aimed at talent recruitment would be based on a bottom-up approach. In the first phase, the cooperation will be limited to the cooperation between the Departments of Genetics and Biochemistry of the FNS and the School for exceptionally gifted children in Bratislava. This phase will represent a trial version for future activity. In the second phase, additional Departments of the FNS as well as high schools in Bratislava will be integrated and in the third and forth phase the project will expand to Bratislava- and nation-wide, respectively.

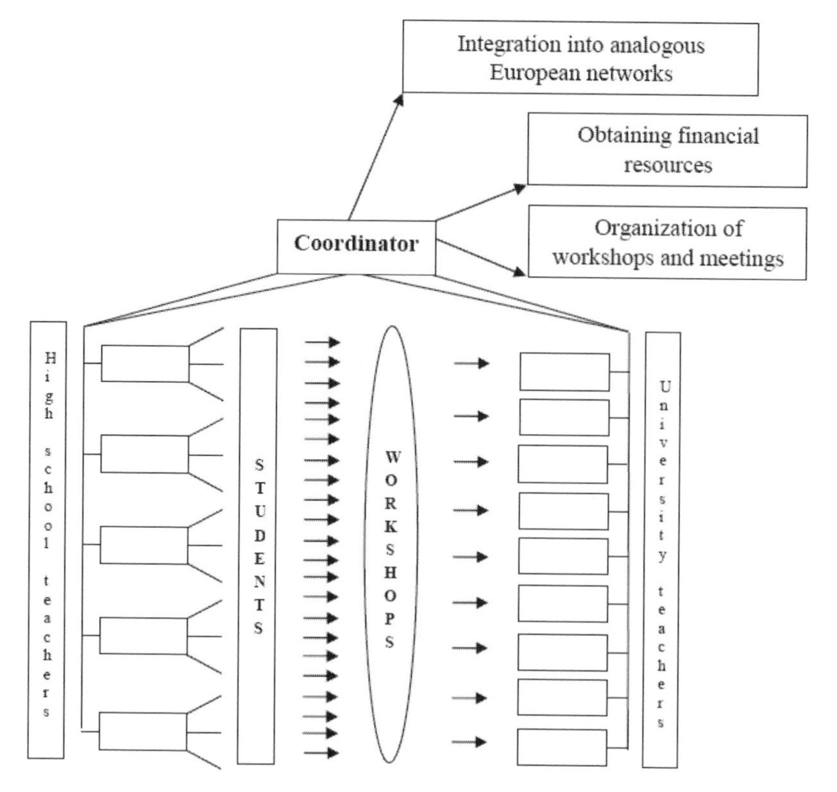


**Figure 1.** Scheme of the perspective nation-wide network for recruitment of talents for biomedical sciences. The coordinator (not necessarily one person) will interact with high school students through teachers or via internet ads. The students will be provided with a database of university teachers, their laboratories, list of projects and workshops. The workshops will serve as the main sites for selecting students for work in a particular laboratory.

## 5. Final remarks

The only Slovak governmental grant agency supporting educational activities rejected the proposal for building the talent network based on a vague statement and without any (positive or negative) feedback. However, the potential fruits of the effort are too irresistible to give up yet. Exchange of ideas with communities having long-term experience with successful quest for talents should be instrumental in overcoming the current (and perhaps transient) obstacles in Slovakia.

## Acknowledgments

I wish to thank Ladislav Kováč for many inspirations and helpful discussions and Jozef Nosek for comments and sharing his ideas.

**References:**

[1] L. Kováč, Charvát's question. *Vesmir* 87 (2004), 47-49. [In Slovak]
[2] J. Mervis, Down for the count? *Science* 300 (2003) 1070–1074.
[3] P. Csermely, Recruiting the younger generation to science. *EMBO Reports* 4 (2003) 825-828.
[4] M. Hejný and M. Koman, Samples of Problem Nets. (Creative Approach to Teaching-Learning Situations). 1993.
[5] J. Laznibatová, Gifted child, its development, education and support. Bratislava, Iris, 2001. [In Slovak]
[6] Making the right moves: A practical guide to scientific management for postdocs and new faculty. Based on the Burroughs Wellcome Fund & Howard Hughes Medical Institute Course in scientific management for the beginning academic investigators, 2004.

**Websites:**
http://www.nadanedeti.sk/menu.php
http://www.pedf.cuni.cz/kmdm/katedra/pracovnici/hejny.htm

*Science Education: Best Practices of Research Training for Students under 21*
*P. Csermely et al. (Eds.)*
*IOS Press, 2005*


# Concluding Remarks and Perspectives

**Peter CSERMELY**
*Hungarian Student Research Foundation*
*H-1146 Budapest, Ajtósi-Dürer sor 19-21*
*csermely@puskin.sote.hu*

Science education is an important element of the recruitment of further generations for scientific research. In this complex process a key point is the science education of high school students, who are in a very susceptible age to ask clear questions about the world around them, and to seek answers in a methodological way, as science does. There are numerous important points to make a high-school oriented project successful. Some of them:

- it is good, if the selection is based on long-term success and not one-time achievement or property (like IQ or SAT tests) and includes motivation as a major part
- it helps the selection if the project gives no benefits (stipends, bonus points for university or other studies) but the joy of scientific research; it is important, however, that the project should be either free, or should provide and easy and not humiliating process to ask for financial support for a full coverage of the project fees if needed
- the project has to reflect "real life" with all its successes and failures, it should show the open-ended, unlimited possibilities of scientific research
- a continued and personalized support is important to make the students' commitment strong and long-lasting (the personal presence of top scientists is important: they may serve as important role models)
- the project has to be hierarchy-free, democratic, where students are not "kids" but treated as young scientists, as equal peers; it is very helpful if mixed student/teacher research teams can be made; trust in the students and in their future achievements is a must for their success
- it is helpful if the style is playful and humorous
- regular occasions for community support (discussions, clubs, conferences, excursions, etc.) should be arranged
- it is good of the support is extended to communication skills, leadership skills and strengthens emotional background to cope with the frustrations and personal problems
- the build-up of an acquaintance network between young scientists themselves, young scientists and their teachers, mentors, as well as persons from the "outside life" (industry, political life and other sectors of the society) is highly helpful
- most of the above features can be strengthened and achieved if the initiative is led by the research students themselves.

The workshop spurred a great interest in the 18 participating NATO and partner countries and gave a comprehensive survey of existing, highly successful examples of scientific research training in Europe, in the USA, Middle East and in Asia. The concentrated introduction of the best practices provided a unique opportunity to learn successful elements from other initiatives as well as to implement these techniques to other Central-Eastern European (mostly NATO Partner) countries.

The workshop made it clear that the talent-support/science recruitment practices have a great diversity, which is a big asset of the group. However, this also requires a careful structuring of the organisation fostering continuous collaboration. The diversity prevails in:

- Target group (talented, underprivileged, motivated students; science teachers; society around the students, their parents, family, peers, neighbours)
- Content (subject-based projects; participation in top and real science; broad, exciting projects; fun-type projects)
- Methods (courses, summer schools, weekend seminars, continuous projects, lap-type projects, school-type projects, computer-based projects, cyber-courses, distance-learning)
- Aims (information transfer, ability development, raising self-confidence, raising long-lasting interest and commitment to science, help in science communication, help in applied research, promotion of public understanding)

The participants of the meeting adopted the following agenda to foster future world-wide collaboration between scientific research training initiatives:

1. The participants have decided to establish a Network of Youth Excellence and have set up a preparatory committee to make a draft of Memorandum of Understanding, ask for nominations for the Board of the Network, organise the elections and help the actual work to be started. The Network will
   - promote cooperation between existing scientific research training projects for talented high school students
   - promote research collaborations between students and teachers of different programs and countries
   - facilitate the collaboration with international organizations of young scientists such as the World Academy of Young Scientists (WAYS)
   - better the existing projects by exchanging their experiences and outlining successful orgnanizational and fundraising tactics
   - help the initiation of scientific research training projects in countries where they currently do not exist (starting in Slovakia, Romania and Malawi)
   - initiate international joint scientific student/teacher projects
   - promote the participation of students in the organization of research training programs
   - encourage a dialog on the ethical and responsible conduct of research and use of scientific knowledge
   - draw the attention of policy makers and the media to the importance to start the recruitment to scientific research at a very early age.
2. The hosting organisation (www.kutdiak.hu) will develop an interactive web-forum of the Network of Youth Excellence and link the current web-site (**http://www.chaperone.sote.hu/nato.htm**) of the meeting to it. Both the new web-site and an email-network will be used for the dissemination of new practices as well as for the development of multilateral contacts.

3. Participants proposed a follow-up meeting two years from now, in the spring of 2006. As a result of the intensified networking during the coming two years, the follow-up meeting would be an excellent occasion for the evaluation of Network activities and to propose a "know-how" to help the further dissemination of successful practices.

# Author Index